The Human Right to Housing in the Face of Land Policy and Social Citizenship

Michael Kolocek

The Human Right to Housing in the Face of Land Policy and Social Citizenship

A Global Discourse Analysis

palgrave
macmillan

Michael Kolocek
School of Spatial Planning
TU Dortmund University
Dortmund, Germany

ISBN 978-3-319-53488-6 ISBN 978-3-319-53489-3 (eBook)
DOI 10.1007/978-3-319-53489-3

Library of Congress Control Number: 2017937899

Cover illustration: Ron Bambridge / Getty

Printed on acid-free paper

This Palgrave Macmillan imprint is published by Springer Nature
The registered company is Springer International Publishing AG
The registered company address is: Gewerbestrasse 11, 6330 Cham, Switzerland

Acknowledgments

The present book is rooted in the interdisciplinary research project FLOOR (Financial Assistance, Land Policy, and Global Social Rights, www.floorgroup.de), partly funded by the German Research Foundation (Deutsche Forschungsgemeinschaft). FLOOR was separated into three sub-projects that were investigated by Ulrike Davy (University of Bielefeld, FLOOR A), Lutz Leisering (University of Bielefeld, FLOOR B), and Benjamin Davy (TU Dortmund University, FLOOR C). Between 2007 and 2015, FLOOR examined human rights, citizenship, and global social policies from different perspectives (e.g., Buschmann 2013; B. Davy et al. 2013; von Gliszczynski 2015). My research was part of sub-project FLOOR C, socio-ecological land policy. FLOOR C focused on the relationship between the poor and the land (B. Davy 2009).

The empirical database (States Parties reports and Concluding Observations under the monitoring system of the International Covenant on Economic, Social and Cultural Rights) of this book has been kindly provided by Ulrike Davy and her team from the Chair of Constitutional and Administrative Law, International and German Social Law, and Comparative Law (University of Bielefeld). I owe many thanks to Ulrike Davy for providing the data, but also for sharing the passion for the ICESCR discourse as a source for global comparative research.

The principal investigators and the research assistants of the FLOOR project also shared their ideas and critical thoughts. I enjoyed being part of this research group. I want to thank Lutz Leisering for the helpful critiques and comments at decisive points in the course of my research.

Some of my ideas and theories were tested in the annual conferences of the International Academic Association on Planning, Law, and Property Rights (IA PLPR) in Belfast (2012), Haifa (2014), Volos (2015), and Bern (2016). I want to particularly thank Thomas Hartmann and Rachelle Alterman for their helpful comments at different stages of my research.

I am grateful to Benjamin Davy for inventing the diffusion maps which show certain findings of the present analysis. I also want to thank Gabi Hartmann, Heinz Kobs, and Nadine Preuß who helped at different stages during the design process of these maps.

The lists of references have been generated with the help of Citavi, a reference management program (www.citavi.com). The list of the States Parties reports and the list of the Concluding Observations of the Committee are based on a Citavi file that Matthias Barutowicz chiefly created as part of the FLOOR C project. Dziękuję bardzo!

I am grateful to Sepideh Abaii, Sattwick Dey Biswas, Jackline Kabahinda, Melanie Halfter, Astrid Maurer, Susanne Syska, and Yitu Yang for asking difficult, but helpful questions. Special thanks go to Luise Buschmann and Kathrina Völkner for their fruitful comments on an early version of this book.

I want to particularly thank Benjamin Davy for the many hundreds of hours of discussion on land, poverty, housing, discourse theory, rights, informality, and many other topics. I learned more than I could put into one single book.

In the present book, I regard 'family' as one actor group of others in the context of human right to housing. In fact, family is much more than that. I am so blessed to have you all in my life. Thank you!

Contents

List of Abbreviations

1951 Convention	Convention Relating to the Status of Refugees
ACHPR	African Charter on Human and Peoples' Rights
ACHR	American Convention on Human Rights
AP-ESC	Additional Protocol to the European Social Charter Providing for a System of Collective Complaints
Arab CHR	Arab Charter of Human Rights
ATLAS.ti	ATLAS = (German) Archiv für Technik, Lebenswelt, Alltagssprache; ti. = text interpretation
BVerfGE	German: Entscheidungen des Bundesverfassungsgerichts (Decisions of the Federal Constitutional Court)
CDA	Critical Discourse Analysis
CEDAW	Convention on the Elimination of All Forms of Discrimination against Women
CERD	Convention on the Elimination of All Forms of Racial Discrimination
CESCR, The Committee	(UN) Committee on Economic, Social and Cultural Rights
CFR	Charter of Fundamental Rights of the European Union
CHR	Commission on Human Rights
CIDA	Canadian International Development Agency
COs	Concluding Observations
COHRE	Centre on Housing Rights and Evictions

CRC	Convention on the Rights of the Child
CRPD	Convention on the Rights of Persons with Disabilities
DALE	Disability-adjusted life expectancy
ECHP	European Community Household Panel
ECHR	European Convention for the Protection of Human Rights and Fundamental Freedoms
ECOSOC	Economic and Social Council
ECSR	European Committee on Social Rights
ECtHR	European Court of Human Rights
e.g.	Latin: Exempli gratia (For Example)
ESAP	Economic Structural Adjustment Programme
ESC	European Social Charta
ESC Rights	Economic, Social, and Cultural Rights
EUROFOUND	European Foundation for the Improvement of Living and Working Conditions
FEANTSA	European Federation of National Organisations Working with the Homeless
FIDH	International Federation for Human Rights
FLOOR	Research Project: Financial Assistance, Land Policy, and Global Social Rights
FLOOR A	Subproject of FLOOR: Social Rights—Towards a Global Human Right to a Civic Minimum
FLOOR B	Subproject of FLOOR: Social Cash Transfers—The Global Construction and Diffusion of the Right to a Monetary Minimum
FLOOR C	Subproject of FLOOR: Socio-ecological Land Policy
FEANTSA	European Federation of National Organisations Working with the Homeless
GATT	General Agreement on Tariffs and Trade
GLTN	Global Land Tool Network
HDI	Human Development Index
HU	Hermeneutic Unit
IH	Inadequate Housing (used in tables)
ICCPR	International Covenant on Civil and Political Rights
ICESCR	International Covenant on Economic, Social and Cultural Rights
ILO	International Labour Organization
IMF	International Monetary Fund

MDGs	Millennium Development Goals
MWC	Convention on the Protection of the Rights of All Migrant Workers and Members of Their Families
NPE	New Political Economy
OECD	Organization for Economic Co-operation and Development
oc	Operational category
OP-ICESCR	Optional Protocol to the International Covenant on Economic, Social and Cultural Rights
p.	Page
Para.	Paragraph(s)
PD	Primary Document
RESC	Revised European Social Charta
SAP	Structural Adjustment Program
SKAD	Sociology of Knowledge Approach to Discourse
SPIH	Spaces of Inadequate Housing (used in tables and maps)
SPR	States Parties Reports
SRZ	Stadt- und Regionwissenschaftliches Zentrum
UDHR	Universal Declaration of Human Rights
UN	United Nations
UNCHS	United Nations Center for Human Settlements (Habitat)
UNDP	United Nations Development Program
UN-HABITAT	United Nations Human Settlements Programme
UNHRP	United Nations Housing Rights Programme
UNESCO	United Nations Scientific and Cultural Organization
USAID	United States Agency for International Development
VerpackV	German: Verordnung über die Vermeidung und Verwertung von Verpackungsabfällen (Regulation on the recovery and avoidence of packaging waste)
WHO	World Health Organization
WTO	World Trade Organization

List of States' Abbreviations

AFG	Afghanistan
ARG	Argentina
AUT	Austria
BEL	Belgium
BIH	Bosnia and Herzegovina
BOL	Bolivia
BRA	Brazil
CHN	People's Republic of China
CMR	Cameroon
COD	Democratic Republic of the Congo
COL	Colombia
CZE	Czech Republic
DOM	Dominican Republic
EGY	Egypt
FIN	Finland
GEO	Georgia
GIN	Guinea
GMB	The Gambia
GRC	Greece
HND	Honduras
HRV	Croatia
ISR	Israel
ITA	Italy

KHM	Cambodia
KOR	Republic of Korea
LKA	Sri Lanka
MAR	Morocco
MCO	Monaco
MDG	Madagascar
MEX	Mexico
MKD	The former Yugoslav Republic of Macedonia
NIC	Nicaragua
PAN	Panama
PHL	The Philippines
POL	Poland
PRY	Paraguay
SLB	Solomon Islands
SMR	San Marino
SRB	Serbia
TCD	Chad
USA	United States of America

List of Figures

List of Tables

List of Diffusion Maps

1

Introduction

Housing is a human right. The present book is a story about the content of this right. It brings the different interpretations of housing rights to the center of attention.

> The national housing indicators for 2010 show that 3,573,653 people are homeless, comprising 11.2% of the approximated 31,981,536 population in 2010. (Uganda 2012: para. 120)

> Homelessness does not exist in Liechtenstein. A facility for homeless persons had to be closed for lack of use. (Liechtenstein 2004: para. 167)

> [W]e believe that our approach is the right one. The people who live in bed-space apartments do so largely out of choice for convenience, economic and other personal reasons. There is ample provision in government hostels—which are clean, safe and well managed—to accommodate the relatively small numbers involved. Yet they remain underutilized, probably because their targeted occupants place a higher value on location and convenience than on the standard of their living environment. (People's Republic of China 2003: para. 592)

> [C]ollective reception centres for asylum-seekers do not offer permanent accommodation but merely meet the need for temporary shelter.

M. Kolocek, *The Human Right to Housing in the Face of Land Policy and Social Citizenship*, DOI 10.1007/978-3-319-53489-3_1

> Consequently, asylum-seekers must accept the inconveniences which typically come with a stay in collective reception centres. This point has been confirmed by German constitutional case law. (Germany 2000: para. 195)

The list of housing quotations could be continued for many pages, but let me start at the beginning. On 10 December 1948, after analyzing the Second World War, the United Nations General Assembly adopted the Universal Declaration of Human Rights (UDHR). According to the UDHR, housing is a component of the human right to an adequate standard of living:

> Everyone has the right to a standard of living adequate for the health and well-being of himself and of his family, including food, clothing, housing and medical care and necessary social services, and the right to security in the event of unemployment, sickness, disability, widowhood, old age or other lack of livelihood in circumstances beyond his control. (Article 25, para. 1, UDHR)

The UDHR could be regarded as a global consensus (U. Davy 2015) of all countries worldwide. States can implement and interpret human rights in different ways. Scholars, Nongovernmental organizations (NGOs), or other experts have explored how states have implemented human rights on a policy level. 'Human rights imply the obligation of society to satisfy [. . .] claims. The state must develop institutions and procedures, must plan, must mobilize resources as necessary to meet those claims' (Henkin 1990: 3). Housing is, of course, a prominent spatial planning issue. However, this book does not address the implementation of the human right to housing on a spatial level, but rather on a discourse level. Its purpose is not to find out whether China, Germany, Liechtenstein, Uganda, and more than one hundred other states worldwide have fulfilled their obligation to guarantee housing or not—but how the states talked about said obligation.

The UDHR has no binding character, but was understood as something that would be followed up by legally binding conventions (Eide 2010: 168). The two most prominent treaties implementing the UDHR are the International Covenant on Civil and Political Rights (ICCPR) and the ICESCR. Both covenants expand and define the terms of the Declaration and establish legal obligations to which the member states commit themselves (Craven 2002: 7). The human right to housing is rooted in Article 11 ICESCR. The monitoring mechanism of the ICESCR works in the form of a reporting procedure. Each member state has to submit regular reports in which the state describes its policies for implementing economic, social, and cultural rights (hereafter: ESC rights). A human rights organ, namely the Committee on Economic, Social and Cultural Rights (CESCR) (hereafter: The Committee), responds to the States Parties reports with Concluding Observations. Between 1977 and 2015, the ICESCR member states submitted States Parties reports to which the Committee drew up Concluding Observations. The monitoring system under the ICESCR offers a valuable data source to examine the consideration of the human right to housing on a global level. These States Parties reports and the Concluding Observations are the empirical heart of the present research that answers the following two questions: How did the ICESCR member states interpret their obligations concerning the human right to housing? How did the Committee on Economic, Social and Cultural Rights accompany and influence the ICESCR member states' reporting?

The research is based on the assumption that housing is an important pillar of social policy. Social policy research often takes T.H. Marshall's (1950) essay 'Citizenship and Social Class' as a point of departure. Marshall discussed the development of citizenship with respect to the relationship between (in)equality and welfare. Inequality of wealth is the price for equality of status (in the form of rights), he stated. Social citizenship currently stands for a policy that reaches equality goals and regards all people as members of society with social rights. Deacon (2007 and 2010) examined the relationship between globalization and social policy. He stated that, since about the 1980s, social policy has been undergoing a process of globalization (globalization of social policy); at the same time, global politics have been characterized by a process of socialization (socialization of global politics) (Deacon 2007: 3). T.H. Marshall and Deacon inspired academics to ask if there is such a thing as global social citizenship and what it could mean (B. Davy et al. 2013).

This book will discuss whether there is something such as global social citizenship in the face of the human right to housing and what it means from the perspective of the ICESCR member states.

While Marshall had the states' responsibilities in mind, the current role of the state is often regarded in the context of other actors. Esping-Andersen (2011 [1990][1]) distinguished three ideal types of welfare states based on three different indicators: the rate of de-commodification, the effects upon stratification, and the role of the three stakeholder groups state, market, and family in the field of social policy. Gough (2004a) added the actor group community and emphasized the importance of informal arrangements by observing welfare policies in developing countries. The high interest in the (changing) responsibilities of different actors is one of the most important subjects in both housing as well as social policy discourses. The book will prove that the global attention to the human right to housing has continuously increased during the last four decades. However, during the same time period, the number of people affected by inadequate housing has continuously increased as well. Finally, policymakers all around the world have to accept that they will not be able to completely fulfill the human right to housing. This, however, means that they have to start or keep on thinking about alternative solutions. For such solutions, planning authorities as well as planning scholars should consider the plural meanings of land.

Chapter 15 of Karl Polanyi's famous book 'The Great Transformation' begins with a remarkable statement with respect to land: 'What we call land is an element of nature inextricably interwoven with man's institutions. To isolate it and form a market for it was perhaps the weirdest of all undertakings of our ancestors' (Polanyi 2001 [1944]: 187). Polanyi discussed the relationship between land and labor. He stated that the economic function of land is only one of many others:

> [Land] invests man's life with stability; it is the site of its habitation; it is a condition of physical safety; it is the landscape and the seasons. We might as well imagine his being born without hands and feet as carrying on his

[1] First published in 1990.

> life without land. And yet to separate land from man and to organize society in such a way as to satisfy the requirements of a real-estate market was a vital part of the utopian concept of a market economy. (Polanyi 2001: 187)

The author was right when he emphasized that land is more than just a commodity. As a research discipline, land policy observes the allocation of land rights and the distribution of the gains and losses of land uses (B. Davy 2005). While allocation is about efficiency, distribution is closely related to justice. An appropriate example that demonstrates how land policy shall work as social policy is Hernando de Soto's (2000) land titling approach. De Soto regarded informal working and housing structures in the global South and stated that, through land titling, the 'dead capital' in informal settlements could be woken up. People living in illegal or informal settings shall become members of the market, and consequently, members of the society through rights—that is, property rights. The author propagated the commodification of land as a form of social policy. Scholars often criticized land titling as a neo-liberal approach that would not work in countries of the global South (e.g., Campbell 2013; Davis 2006; Gilbert 2002; Payne 2001) and propagated more cautious concepts to respond to informality (e.g., Roy 2004 and 2005; Neuwirth 2006 and 2011). The present research fits into this debate and discusses the significance of property and land use rights in the context of the human right to housing. Based on a discussion of land in the context of (global) social citizenship, I offer a theory of de-commodification of land use for people affected by inadequate housing.

The book consists of the introduction and two main parts. The first part is a discussion of the main topics and concepts of the book: human rights, discourse analysis, housing, land policy, global social citizenship, and de-commodification of land use. In Chapter 2, I focus on the monitoring system of human rights. The book is not a policy investigation (on the spatial level), but rather puts the content of the human right to housing on a discursive level in the center of attention. The book's findings are based on the use of different tools and terms offered by discourse analysis, which is a research perspective that concentrates on the empirical investigations of discourse (Keller 2013: 3). So far, no attempt to analyze the content of the human right to housing on a global level has been made. Therefore, it is necessary to introduce the concept of discourse analysis in more detail.

I then present the ICESCR with its roots as well as its standing in human rights discourses, including the main speakers (the States Parties and the Committee), actors, and documents involved. The third chapter discusses housing from different points of view. First, I discuss housing through the lens of jurisprudence. Second, I give an overview of housing investigations that are based on discourse analysis. Third, I show that Western scholars' housing research previously focused largely on the tenure question. After that, I present two main forms of inadequate housing: homelessness and Spaces of Inadequate Housing. The fourth chapter evaluates land policy and global social citizenship which is a concept to examine social policy from a global point of view. At the end of this chapter, I develop the theory of de-commodification of land use. De-commodification of land use for people affected by inadequate housing stands for a policy that captures different rights and needs of the affected persons and looks for solutions outside of formal housing and property markets.

The second part of the book answers the research questions presented above. Chapter 5 illustrates in detail how the ICESCR discourse on housing has been changing over the last four decades. In Chapter 6, I give a statistical overview of the global attention to housing subjects. The chapter illustrates the main differences between the states' attention to certain forms of inadequate housing, disadvantaged and vulnerable groups, and actors. Moreover, I point out the main differences and overlaps between the ICESCR member states. While some few states turned on the road to (global) social citizenship, others had certain focal points or remained silent. The book closes with a critical evaluation of the findings in Chapter 7.

Part I

Housing, Rights, Land Policy, and Global Social Citizenship

2

The Monitoring of Human Rights

This chapter first introduces discourse analysis as the methodological concept for the present research aim, which is to examine the monitoring of the implementation of the human right to housing (2.1). Section 2.2 presents the context, the main contents, and the standing of the ICESCR in the human rights discourse. I draw attention to the significance of the monitoring system of the ICESCR for examining the view on housing in a global perspective. The research puts the conversation of both the ICESCR member states and the Committee in the center of interest. These speakers, including the textual apparatus that defines the rules of speaking about ESC rights, will be presented in more detail in Sections 2.3 and 2.4. At the end of this chapter, I point out why discourse analysis is an appropriate concept to explore the view on housing in a global context and why the monitoring system under the ICESCR offers the right data source for such an intention (2.5).

2.1 Discourse Analysis

Discourse analysis stands for a broad catalog of research approaches such as Discourse Analysis, Critical Discourse Analysis, Discourse Linguistics, Linguistic-Historical Analysis of Discourse, Cultural Discourse Research,

M. Kolocek, *The Human Right to Housing in the Face of Land Policy and Social Citizenship*, DOI 10.1007/978-3-319-53489-3_2

and many more (Keller 2013: 13–68). Scholars can analyze any possible topic through discourse analysis, including housing (e.g., Dodson 2007; Fopp 2009; Forte 2002; Kolocek 2012, 2013, 2014 and 2015; Meert et al. 2004; Torck 2001).

Van Dijk acknowledged that the notion of the term is 'fuzzy' (van Dijk 1997: 1). Handbooks and introductions about discourse research (e.g., Keller 2011a and 2013; Keller et al. 2006 and 2008; van Dijk 1997a and 1997b) have the following structure in common: They begin with a discussion of the term discourse, present the roots of discourse research, emphasize its multidisciplinary character, and point out that each research approach is individual. Discourses are, inter alia, 'attempts to stabilize [. . .] attributions of meaning and orders of interpretation and thereby to institutionalize a collectively binding order of knowledge in a social ensemble' (Keller 2013: 2) or, in other words, 'regulated, structural practices of sign usage' (Keller 2011c: 51).

As a first step, it is necessary to distinguish between discourse analysis and discourse theory. Discourse theories are concerned with the development of theoretical perspectives on the linguistic constitution and the meaningfulness of society. Michel Foucault, Emile Durkheim, Pierre Bourdieu, Jürgen Habermas, and many scholars from Cultural Studies, Feminism Theory, and Post-colonialism had an enormous influence on discourse theory (Keller 2013). Discourse analysis concentrates on the empirical investigations of discourse (Keller 2013: 3). Both discourse theory and discourse analysis:

- are concerned with the actual use of (written or spoken) language and other symbolic forms in social practices;
- emphasize that in the practical use of signs, meanings of phenomena are socially constructed, and these phenomena are thereby constituted in their social reality;
- claim that individual instances of interpretation may be understood as parts of a more comprehensive discourse structure that is temporarily produced and stabilized by specific institutional-organizational contexts; and

- assume that the use of symbolic orders is subject to rules of interpretation and action that may be reconstructed. (Keller 2013: 3)

The concept of discourse involves language use, cognition, and interaction in their respective socio-cultural context:

> [D]iscourse analytical studies distinguish various levels, units or constructs within each of these dimensions, and formulate the rules and strategies of their normative or actual uses. [. . .] Discourse analysis thus moves from macro to micro levels of talk, text, and society, and vice versa. (Van Dijk 1997: 32)

The different disciplines that concern themselves with discourse analysis differ by their focal point. The 'recent boom in discourse research' (Keller 2013: 1) has at least two reasons: the exponential growth in the production of knowledge and the enormous expansion in professionalized communication processes and technologies (Keller 2013: 4). Discourse researchers, like other researchers, try to capture this exponential growth of knowledge production. Ironically, by so doing, they themselves produce new knowledge that could be subject to further discourse research.

Discourse analysis aims to ask and answer questions. The countless approaches differ in particular in the type of questions they emphasize. Some scholars concentrate on the statements of speakers while others focus on the speakers themselves. Other scholars explain the rules underlying a discourse. Typical questions in discourse research are as follows: Who is speaking? Who is not? What is the speaker talking about? When? Where? Under what type of circumstances? How? How often? Why? Some studies analyzed the use of language in detail and took every written word or the grammar used seriously (Gill & Whedbee 1997; Cumming & Ono 1997). Other studies also focused on single terms. They observed the relation to other terms in a quantitative manner, often supported with special computer software (Dzudzek et al. 2009; Mattissek 2008). Some investigations were more focused on rules of formation and the practices of discourse production (Fopp 2009) and answered questions such as these: Why does speaker X speak and not speaker Y? Why does the speaker speak in this moment and not earlier or later? What power resources does the speaker have in comparison with another speaker? Who decides what is true? Or, as Foucault stated: '[According to what] rules has a particular statement been made, and consequently according to what rules could other similar statements be made? [H]ow is it that one particular statement appeared rather than another?' (Foucault 2002 [1969]: 30).

Considering housing, one can distinguish easily between questions that planning and policy scholars would raise as compared to discourse analytical questions such as 'How are words and things linked in the housing policy discourse?' (Dodson 2007: 39). Or, as Fopp asked by examining the discussion about homelessness:

> How extensive and prevalent is the use of metaphors in research into housing and homelessness? What social function do they perform? Do they contribute to or reinforce the dominant discourses and metanarratives about homelessness and, if so, how? What is conveyed about people who are homeless when such figures of speech are used about them and their circumstances? Do the metaphors reflect the stated experience of people who are homeless? (Fopp 2009: 273)

Scholars usually combine qualitative and quantitative elements with each other. Sometimes, quantitatively orientated approaches, such as simple word counts generated with the help of computer software, are a suitable starting point for qualitative approaches (Dzudzek et al. 2009).

Keller's (2013) introduction to countless examples presents a good starting orientation about the scope of discourse research. His basic terminology of discourse research (Keller 2013: 72–73) illustrates the main terms used in discourse analysis research and furthermore reminds the researcher to develop his or her own terminology that complements his or her research purpose.

Keller (2009 and 2011b) investigated the public discourses on domestic waste in France and Germany between 1970 and 1995 and discussed the similarities and the differences between both countries. The subsequent reflections about this research process led to later publications about his 'Sociology of Knowledge Approach to Discourse' known as

SKAD (Keller 2009: 10). SKAD has its origin in the sociology of knowledge theory of Berger and Luckmann (1967) and builds a bridge between Culturalist approaches to discourse research and Michel Foucault's discourse theories (Keller 2013: 61–135 and 2011c). Keller (2013: 54 and 2011c: 52) responded to a key critical argument against Foucault that the philosopher did not give any theoretically or methodologically and consistently developed proposal for the conduct of discourse. In Keller's own words:

> [SKAD] gives priority to Berger and Luckmann, because they establish a dialectical perspective on society as 'objective reality' and as 'subjective reality', becoming 'real' through all kinds of knowledge. It uses Foucault's ideas and concepts [. . .] in order to explore in more detail the processes of institutionalization and transformation of symbolic ordering neglected by Berger and Luckmann tradition, and it refers to methodology and methods of qualitative (interpretative) inquiry, close to the perspective of the latter. (Keller 2011c: 48)

The author stated that sociology, with its methodological richness, allows a broader underpinning of discourse research than is possible in approaches that have their roots in linguistics or discourse theory (Keller 2013: 64). In contrast to other approaches, SKAD has some advantages, inter alia its high level of transparency and its ability to self-correct (Keller 2011b: 11), which is based on the Grounded Theory (Glaser & Strauss 2008 [1967]). Keller (2011c and 2013) developed many research categories that play a certain role in different phases of the research process. Examples for terms that appear in discourse research are categories, codes, meaning patterns (in German: Deutungsmuster), dispositif (dispositive, Keller used the French term), phenomenal structure (in German: Phänomenstruktur), or discursive events, storyline, narratives, and many more.

By analyzing the social construction of waste, Keller (2009: 55–60) observed in both countries, Germany and France, a vast amount of data: articles from daily and weekly newspapers, legal texts, advertisements, flyers, interviews with experts, and many more. Keller (2009: 60–61) started his research process by ordering data and then systematically reduced the data for a more detailed analysis. By doing so, Keller read every text document (discourse fragment) reiteratively and wrote summarizing memos, sorted the discourse fragments with matrices, and developed different codes, first for each text segment, then for each discourse fragment as a whole.

Keller (2013: 73) understood dispositif to mean the material and ideational infrastructure, for example, the laws, codes of behavior, buildings, and measuring devices. Dispositif is often translated as 'apparatus' (Keller 2011c: 47). Keller differentiated between dispositifs of discourse production and dispositifs of infrastructure that emerge out of a discourse:

> A dispositif is both: the institutional foundation, the total of all material, practical, personal, cognitive, and normative infrastructure of discourse production, and also the infrastructures of implementation emerging out of discursively configured problematizations of fields of practice. [. . .] Dispositifs mediate between discourses and fields of practice. (Keller 2011c: 56)

The author claimed that '[a] dispositif may also include the formal, institutionally legitimized texts that indicate how to proceed in specific cases of application (such as laws, regulations, legal prescriptions and so on)' (Keller 2013: 71). For the interpretative reconstruction of discourses, Keller proposed a quite complex interpretative repertoire (Keller 2011c: 57). This repertoire will not be presented here in its full scope, but only to the extent that is necessary for the present research. Keller (2013: 112–127) proposed three different concepts (tools): phenomenal or problem structures, meaning patterns, and narrative structure. Phenomenal structures are used to put the different elements (codes) in a discourse into a sort of structure. This means, for instance, in the discourse of a public problem, 'determining the nature of the problem [. . .], designating the causal relationships (cause-effect), responsibilities, problem-dimensions, value-implications, moral and aesthetic values, consequences, action possibilities and so on' (Keller 2013: 115). Meaning patterns are, according to Keller (2013: 113), interpretative schemes, or frames that link 'various sense making elements into a coherent (not necessarily consistent) meaning-figure that may appear in various manifestations' (Keller 2013: 122). Keller (2013: 123–124) developed the technology-meaning pattern risk in the German debate on waste. He stressed that it is the researcher who labels the meaning pattern. The third tool is the narrative structure (Keller 2013: 124–127). The concept of narrative structure (or story line) helps the researcher to 'tell the story' by working out the plot and narrative patterns or dividing the discourse into different time periods (episodes).

In his exploration of the waste discourse, Keller came to the conclusion that the German discourse on waste disposal consisted of two antagonistic discourses: A structurally conservative discourse that propagated technical modernization through the market, and a culturally critical discourse that stated political-ecological restructuring through the socialization of the waste problem (Keller 2009: 308–309). In contrast, France was characterized through a hegemonic administrative discourse, which means that the discourse emphasized the state representation of collective national interests (Keller 2009: 309). A linguistic perspective is in Keller's approach of secondary importance:

> Questions of grammar, syntax, the use of specific rhetorical devices (metaphor, synecdoche, metonymy and so on) may be of interest in individual cases [but] so long as there is no presumption of a discourse specific relation to grammar, rhetorical elements, contents and speaker positions, these kinds of analytical instruments can largely be dispensed with for purposes of social science research. (Keller 2013: 75–76)

With SKAD, Keller has built a bridge between the text analyses on one side and the analysis of the dispositif and the discourse practices on the other side. In other words: SKAD combines tools from the methodological toolbox of the German Sociology of Knowledge with tools from Foucault's toolbox. While the Sociology of Knowledge has been primarily orientated toward the microanalysis of knowledge, Michel Foucault observed the powerful institutional mechanisms of the circulation of knowledge and the meanings of different practices in this process (Keller 2013: 62–62). Because SKAD is able to combine both, scholars can profit from SKAD by analyzing diverse subjects. As Keller (2013: 67) stated, the usefulness of the term discourse depends on the respective suitability and its justified use for a specific research interest. This also applies to other terms and tools of discourse research. Scholars must develop their own terms and tools.

The present study aims to analyze the content of the human right to housing under the monitoring system of the ICESCR. Table 2.1 shows the discourse definition as well as the terms and tools that are of central relevance for such an approach.

A **discourse** consists of the totality of statements made by **speakers**. The center of attention is the ICESCR discourse from its beginning in 1977 until 2015 and the speakers' way of reporting their implementation of the human right to housing. The book does not examine whether the states successfully solve problems related to housing. Instead, this study concentrates on the states' (and the Committee's) central positions on addressing housing problems. The ICESCR member states and the Committee are not examined as actors, but as speakers. They will be presented and discussed in more detail in Sections 2.3 and 2.4.

The speakers have written down (published) their statements in text documents, or in other words, **discourse fragments**. The discourse fragments are States Parties reports, Concluding Observations and two General Comments. Ulrike Davy and her team from FLOOR A discovered the potential of the ICESCR discourse. They collected the States Parties reports and the Concluding Observations and prepared them for research. FLOOR A examines the contents of economic and social rights (Himpe 2013) and focuses particularly on the rights to social security and social assistance (Article 9 ICESCR), as well as the right to an adequate standard of living (Article 11 ICESCR) (Buschmann 2010 and 2013; U. Davy 2013 and 2014). I am thankful to Ulrike Davy who kindly provided me with her data for my own research. I also thank Luise Buschmann, Ulrike Davy, and Nina-Claire Himpe for their support in comprehending the complexity of the human rights system, including the monitoring mechanism of the ICESCR. The States Parties reports are called Database FLOOR A, SPR and the Concluding Observations are called Database FLOOR A, CO. A State Party report is addressed to the Committee. The

Table 2.1 Terms and tools

Term/Tool	Explanation	In detail
Discourse	ICESCR discourse. Totality of statements made by speakers, written down in discourse fragments.	The discourse on the human right to housing under the implementation mechanism of the ICESCR between 1977 and 2015.
Speakers	Actors that produce discourse fragments in the ICESCR discourse.	ICESCR States Parties The Committee on Economic, Social and Cultural Rights
Discourse fragments	Text documents. Sum of statements. Discourse fragments are addressed to one of the two speakers. States Parties reports are addressed to the Committee. Concluding Observations and General Comments are addressed to the States Parties.	States Parties reports (Database FLOOR A, SPR) Concluding Observations (Database FLOOR A, CO) General Comments No. 4 and 7
Dispositif	Textual apparatus that encompasses the main rules upon which the ICESCR discourse is based.	Articles 17–25 ICESCR Rules of Procedure Reporting Guidelines General Comments No. 1, 2, and 3 Limburg Principles Maastricht Guidelines
Narrative of the discourse	Storyline. The discourse is separated into three time periods (episodes): 1977–1989, 1990–1999, and 2000–2015.	The changing incidence of codes and meaning patterns in different time periods
Discourse event	An event that influenced the narrative of the discourse. Changing of the dispositif.	A state ratifies the ICESCR A state submits a State Party report Establishment of the Committee in 1986 Publication of new textual apparatus (see dispositif)
Quotation	Coded text sequence in the discourse fragment: Word, sentence, paragraph, page, section.	For instance: 'There are no homeless people in our country.'

(*continued*)

Table 2.1 (continued)

Term/Tool	Explanation	In detail
Meaning pattern	Tool for interpretation. Based on the incidence and overlaps of certain quotations in a discourse fragment.	Social citizenship De-commodification of land use
Code	Tool to capture the contents of the statements.	For instance: Inadequate Housing, homelessness, slums, refugees, causes, measures, NGOs, market actors

Committee then writes Concluding Observations addressed to the State Party. Moreover, each year, the Committee publishes General Comments that present its view on a certain right or aspect of the Covenant. Two General Comments (CESCR 1991a and 1997) concern housing. General Comments are somehow both discourse fragments and part of the dispositif because they are addressed to all States Parties, and they are concerned with both the main contents of the Covenant—the economic, social, and cultural rights (ESC rights)—as well as the rules of reporting. The General Comments that address matters of housing are regarded as discourse fragments. General Comments No. 1, 2, and 3 (CESCR 1989, 1990a, and 1990b) are treated as dispositif. As most discourse fragments are accessible on the Internet, the number of potential readers is considerably high.

In a broad understanding, **dispositif** means everything that is not a discourse fragment, but influences the discourse. My understanding of dispositif is a narrower one and close to Keller's description of the dispositif of discourse production. When the speakers produced discourse fragments, they had to follow certain rules. The dispositif encompasses the textual apparatus that defines these rules. Some of the rules can be found in the Covenant (Article 17–25 ICESCR). General Comments No. 1, 2, and 3 (CESCR 1989, 1990a, and 1990b) are also considered dispositive because they emphasize the rules of reporting.

Further documents are the Rules of Procedure (CESCR 1993) and the Reporting Guidelines (CESCR 1986, 1991b, and 2009). These documents have been produced (written) by the Committee itself, which requires some explanation. With respect to 'the power question', the States Parties and the Committee are not equal. In contrast to the States Parties, the Committee can change the rules of speaking, which it has done. The main contents of these documents will be presented in Section 2.3. Furthermore, I will present the Limburg Principles (1987) and the Maastricht Guidelines (1997) in more detail in Section 3.1.

Discourse events change the **narrative of the discourse** when new speakers (new states or the Committee) either join the discourse or when the rules that are underlying the discourse change. Ratification of the ICESCR by a state or the submitting of a report by a State Party is regarded here as a discourse event. The establishment of the Committee in 1986 is perhaps the most influential discourse event. Changes of the dispositif, like publications of new textual apparatus, are discourse events that directly influenced the narrative of the discourse. However, a lot of events also changed the narrative of the discourse indirectly. The fall of the Berlin wall and its territorial consequences in Europe, or the collapse of the Socialist Federal Republic of Yugoslavia, are such events because new states were born that after their independence quickly ratified the ICESCR. As the discourse has been going on for nearly 40 years, one of the most interesting questions is if and to what degree the discussion has changed over the decades. That is why I distinguish between three time periods. This distinction allows for comparisons on a quantitative and qualitative level.

A **quotation** is a coded text segment in the discourse fragment. Depending on the code (see below), this could comprise many paragraphs (in which, for instance, a State Party reports the code homelessness) or be a single word (such as municipality).

A **meaning pattern** is the main tool to interpret the discourse fragments in the context of the research questions. In a technical sense, meaning patterns are statements in discourse fragments that, via the incidence and overlaps of certain quotations, show a special view of the speaker. As the research asks for the incidence of (global) social citizenship and de-commodification of land use in the reporting of the

ICESCR States Parties, these approaches (theories) have been translated into meaning patterns that fit with the states' way of reporting the human right to housing.

The last term suits well to discuss some technical aspects of the present research. In qualitative research, a **code** captures the meaning in the data: 'Codes are used as classification devices at different levels of abstraction in order to create sets of related information units for the purpose of comparison' (Friese 2013: 17). Generating codes is one of the main techniques of discourse research. Codes are usually developed in a process that combines inductive category development and deductive category application (Mayring 2000). The most well-known approach emphasizing that codes, hypothesis, or concepts shall be systematically worked out in relation to the data during the research process (in other words: inductively) is the Grounded Theory (Glaser and Strauß 2008). 'I have no data yet. It is a capital mistake to theorise before one has data. Insensibly one begins to twist facts to suit theories, instead of theories to suit facts' (Sherlock Holmes in the novel 'A Scandal in Bohemia', by Doyle 2003 [1891]: 242). Arthur Conan Doyle could not know anything about Grounded Theory. He nonetheless captured the gist of the theory's philosophy. 'Generating a theory from the data means that most hypotheses and concepts not only come from the data, but are systematically worked out in relation to the data during the course of research' (Glaser and Strauß 2008: 6). Glaser and Strauß (2008) and many other scholars developed several approaches for generating codes—such as substantive coding, open coding, and selective coding (see, e.g., Mey & Mruck 2011); their common ground is that the codes are mainly (but not exclusively) generated inductively. Moreover, Grounded Theory emphasizes that codes and theories are generated in an iterative process.

During the research process, I did not utilize any of the approaches under the umbrella of Grounded Theory as a step-by-step instruction. Considering some aspects, such as the theoretical data sampling (Glaser and Strauß 2008: 45–77), the research even contradicted the philosophy of Grounded Theory. With respect to the codes, however, Grounded Theory influenced the present approach insofar, as the theory emphasizes that the category building (generating of codes) works inductively and involves multiple iterations.

The whole text analysis was done with ATLAS.ti, a computer software for qualitative data analysis (Friese 2013 and 2014; Friese & Ringmayr 2014). The original idea of ATLAS.ti was, from the words of one of its initiators, 'to build an archive of modern everyday culture by compiling a database with all sorts of verbal data from different qualitative research studies that could be used as a resource for secondary analysis' (Legewie 2014: 1). Legewie (2014) reported how Anselm Straus, one of the fathers of Grounded Theory, inspired the development of the software. The idea for ATLAS.ti came up in the late 1980s when Legewie and his team had to analyze 60 in-depth interviews of a longitudinal study:

> What we needed was a software tool that would help qualitative researchers keep their many text documents in check, code or annotate selected text segments and construct semantic networks out of the code lists that emerged in the course of a research project. (Legewie 2014: 2)

The strength of ATLAS.ti is that the software allows for a vast amount of text data—without abandoning the qualitative aspects. Friese emphasized the benefits of a software-supported approach for such a purpose:

> Software changes the way we build up coding systems. The process becomes much more exploratory due to the end ease of renaming and modifying codes. Computers also change the way we ask questions about the data. Data analysis procedures have become much more sophisticated because, for a computer, it is much easier to find things in the data and to output results. Also, [it] makes it easier to combine qualitative and quantitative methods, which of course does not preclude a purely qualitative approach. It allows qualitative researchers to move out of black box analysis and to make the entire analytic process more transparent. (Friese 2014: 3)

For such an 'analytic strategy' (Woolf 2014) ATLAS.ti offers several tools for data management, coding, and writing different kinds of memos to query the coded data, as well as to visualize the research findings (Friese 2014). Using the software powerfully means, as Woolf noticed, to use it 'from [. . .] the start of the data analysis all the way through to the end, fulfilling the needs of every phase while remaining true throughout to the iterative and emergent spirit of qualitative research' (Woolf 2014: 1).

This chapter has introduced discourse analysis from a theoretical and methodological point of view. Table 2.1 (p. 17) gives a sort of abstract overview of the main tools and terms necessary to analyze the discourse on the human right to housing. Of course, Table 2.1 still leaves some room for concretization, which will be filled by the following sections.

2.2 The International Covenant on Economic, Social, and Cultural Rights

The idea of human rights was born from the experiences of the Second World War (Henkin 1990). Shortly after the adoption of the UDHR in 1948, the negotiations to convert the Declaration into a more binding international treaty started. This took 20 years, although it was planned to take only three (Buschmann 2013: 9). As the Declaration contains different categories of rights—civil and political rights on the one hand, and economic, social, and cultural rights on the other hand—it was decided to draft two separate treaties instead of one single document (Craven 2002: 7). In 1966, the ICCPR and the ICESCR were adopted. Because of a slow process of ratification, it took another 10 years for both covenants to go into effect.

The ICESCR is one of nine core international human rights treaties and has been almost exclusively evaluated in legal discourses about human rights (Alston 1992; Buschmann 2010 and 2013; Craven 2002; U. Davy 2013 and 2014; Eide et al. 2001; Rosas 2001; Saul et al. 2014; Young 2008). It consists of a preamble in which the member states point out the importance of the inherent dignity of the human person as well as individuals' duties to other individuals and the community with respect to ESC rights.

The ICESCR is separated into five parts that contain 31 articles overall. Part I contains only Article 1 (self-determination), which is identical with Article 1 ICCPR. The right to self-determination is often cited as one of the cornerstones of the international system, but it is also frequently a concept of confusion and controversy (Rosas 2001: 111). It is not seen as an individual right, but rather as a collective right which, as a consequence, cannot be subject to individual complaints in case of its violation (Rosas 2001: 115). Saul et al. (2014: 41) found that from the early to mid-2000s, the CESCR has also expressly recognized indigenous people under Article 1 in its Concluding Observations. Part II contains Articles 2 to 5. Each State Party of the Covenant commits itself

> to take steps, individually and through international assistance and co-operation, especially economic and technical, to the maximum of its available resources, with a view to achieving progressively the full realization of the rights recognized in the present Covenant by all appropriate means, including particularly the adoption of legislative measures. (Article 2, para. 1, ICESCR)

The phrase 'to the maximum of its available resources' obliges the states 'regardless of the level of economic development to ensure respect for minimum subsistence rights for all' (Limburg Principles 1987: para. 25) and includes 'both the resources within a State and those available from the international community through international cooperation and assistance' (Limburg Principles 1987: para. 26). The states have often argued that the implementation of economic, social, and cultural rights goes beyond their available resources. However, 'resource scarcity does not relieve States of certain minimum obligations in respect of the implementation of economic, social, and cultural rights' (Maastricht Guidelines 1997: para. 10). Article 2, para. 1, can furthermore be interpreted as an obligation to think in a global dimension about the realization of ESC rights. It reminds the states to take their steps not only individually, but also through international assistance and co-operation. In 1990, the Committee published a General Comment that evaluated the meaning of Article 2, para. 1, in more detail (CESCR 1991b).

Article 2, para. 2, stresses that the rights of the Covenant apply to everyone 'without discrimination of any kind'. This includes discrimination 'as a result of the unequal enjoyment of economic, social, and cultural rights' (Limburg Principles 1987: para. 38), and it further demands that the States Parties should 'prohibit private persons and bodies from practising discrimination in any field of public life' (Limburg Principles 1987: para. 40). Article 2, para. 2, could be regarded as an obligation to reach equality in the enjoyment of ESC rights. Article 3 highlights explicitly the equality between men and women. Article 4 stresses that the enjoyment of economic, social, and cultural rights has its limitations, but, as the Limburg Principles substantiate, the article was 'primarily intended to be protective of the rights of individuals rather than permissive of the imposition of limitations by the State' (Limburg Principles 1987: para. 46). Article 5 'underlines the fact that there is no general, implied or residual right for a State to impose limitations beyond those which are specifically provided for in the law' (Limburg Principles 1987: para. 57).

Part III (Articles 6 to 15) is 'the heart of the covenant' (Craven 2002: 22) because it includes the main contents of the Covenant, that is, the ESC rights. They consist of labor rights (Articles 6 to 8), the right to social security (Article 9), family rights (Article 10), the right to health (Article 12), the right to free education (Articles 13 and 14), the right to participation in culture (Article 15), and the right to an adequate standard of living, including the right to housing:

> The States Parties to the present Covenant recognize the right of everyone to an adequate standard of living for himself and his family, including adequate food, clothing and housing, and to the continuous improvement of living conditions. The States Parties will take appropriate steps to ensure the realization of this right, recognizing to this effect the essential importance of international cooperation based on free consent. (Article 11, para.1, ICESCR)

Here again, the states are reminded of international cooperation, although based on free consent. Eide concluded that '[t]he term "adequate standard of living" has not been given a more precise definition in the relevant international instruments [. . .]. The essential point is that everyone shall be able [. . .] to enjoy the basic needs under conditions of dignity' (Eide 2001: 133). Pogge (2007: 90) claimed that the right to an adequate standard of living is one of the most frequently unfulfilled human rights. By analyzing the drafting process with regard to Article 11

ICESCR, Buschmann (2010) stated: Although the debating states were absolutely not in agreement about the contents and meaning of the term inadequate, the debate about an adequate standard of living did not reach the necessary depth; thus indices or benchmarks were not part of the debate (Buschmann 2010: 22–24). This 'fuzziness', however, left room for interpretation in the States Parties reports, the Concluding Observations, and many other sources.

Articles 16–25 (part IV) regulate the system of monitoring. Craven (2001: 458) remarked that the reporting system in part IV 'does not clearly identify which body has central responsibility for supervision (ECOSOC or Commission on Human Rights)' (Craven 2001: 458). The system of supervision works in the form of a reporting procedure and will be presented in more detail during the discussion on the Committee's role (2.3). Part V (Articles 26–31) regulates the modalities of ratification and entry force.

In human rights discourses, ESC rights have, for a long time, been regarded skeptically. The ICESCR does not solely differ from the ICCPR with respect to its rights. In opposition to the ICCPR, the rights of the ICESCR have often not been considered as an individual's rights, but rather as the state's obligations (Henkin 1979). Later, Henkin emphasized important differences between the two covenants, but he also admitted: 'As a matter of law, however, I do not think any of these differences critical. The Convention on Economic, Social, and Cultural Rights uses language of obligation, not merely of aspiration or hope' (Henkin 1990: 33). For a long time, ESC rights had the status of 'second generation rights' (Dean 2007: 2). One reason for the distinction between first and second generation rights was, as Craven (2002: 9) stated, the ideological conflict between East and West during the drafting process of the Covenant. The Soviet States championed the cause of economic, social, and cultural rights, which they associated with the aims of the socialist society, while the Western States asserted the priority of civil and political rights as the foundation of liberty and democracy in the free world. In contrast to Craven's explanation, Eide denied the 'widespread myth' (Eide 2010: 164) that the then Socialist countries introduced social and cultural rights in the negotiations. By discussing the founding process of the UDHR, U. Davy (2013: S21) stated that the Eastern European countries were not the most energetic champions of economic and social rights. Craven (2002: 7–16) discussed the different skepticism with which economic, social, and cultural rights were considered. He stated that 'there are no really convincing arguments either for denying economic, social, and cultural rights the status of human rights or for maintaining absolute distinctions between them and civil and political rights' (Craven 2002: 16).

In contrast to the ICCPR, for a long time, no mechanism for individuals' complaints existed for the ICESCR. Individuals could not claim a violation of their ESC rights rooted in the ICESCR. In 2008, the General Assembly adopted the Optional Protocol to the ICESCR (hereafter: ICESCR-OP 2008). The Protocol came into force in May 2013 when the tenth State ratified or acceded to the treaty (Article 18 ICESCR-OP 2008). Since then, individuals or groups of individuals have been able to claim a violation of their rights to the Committee on Economic, Social and Cultural Rights (Article 2 ICESCR-OP). In 2002, Craven called the ongoing drafting process of the Protocol to be 'one of the most significant initiatives of the Committee' (Craven 2002: x). The Protocol was debated intensively in human rights discourses since the idea and the first drafts were public (e.g., Mahon 2008; Scheinin 2006; Vandenbogaerde & Vandenhole 2010). Fritzsche (2009: 100–101) listed eight signs that attention to ESC rights was on the rise. The Protocol is one of them. Others are the growing attention to poverty, the agenda setting of NGOs and other non-state actors and the emergence of 'new' human rights, such as the right to water (CESCR 2002a). The adoption of the Optional Protocol raised the standing of the ICESCR.

To sum up, human rights scholars nowadays agree that the Covenant is of the same importance as the ICCPR. Nevertheless, only a few experts would deny that the first 10 years (1976–1986) after the ICESCR went into effect were characterized by a high level of ignorance and uncertainty regarding the real contents and, in particular, the standing of ESC rights. However, this has changed considerably since 1986. This year can be regarded as the year of the most influential discourse event: The Committee on Economic, Social and Cultural rights was established.

2.3 The Committee on Economic, Social and Cultural Rights

The Committee is both a speaker in the discourse and the producer of textual apparatus that encompasses the reporting rules. It has the power to influence both the contents of the discourse as well as the dispositif.

The Committee is one of ten treaty bodies that monitor the implementation of the core UN international human rights treaties. The Committee's European counterpart is the European Committee of Social Rights (ECSR) (Alston 1992: 474). The monitoring institution for the ICCPR is the Human Rights Committee (HRC). In contrast to other human rights committees, the Committee is technically only an organ of the UN with the mandate to assist the ECOSOC in consideration of the States Parties reports (Craven 2001: 461).

The Committee was established in 1986, 'almost in desperation as a result of the Working Group's inadequacies' (Alston 1992: 473). Part IV of the Covenant does not clearly state whether the human rights organ ECOSOC or the Commission on Human Rights (CHR) had responsibility for supervision (Craven 2001: 458). The ECOSOC was too large, too unwieldy in organizational terms, and did not have enough resources to undertake the tasks entrusted to the terms of the Covenant, and the CHR had a full timetable, too (Craven 2002: 459). The Committee is a body of 18 independent experts who shall have recognized competence in the field of human rights (ECOSOC 1985: para. b). The main contents of its structure and work and the process of the working sessions are described in 72 Rules of Procedure (CESCR 1993). The Committee should meet annually for a period of up to three weeks (CESCR 1993: rule 1). In 1995, the ECOSOC decided that the Committee should hold another three-week session each year 'in order to cope with its workload' (ECOSOC 1995: para. d). The Committee's basic tasks are to review the reports and reply to each report with Concluding Observations.

The review of the reports works as follows: Five of the Committee members initially consider the submitted report in a pre-session working group with the purpose to get a handle on the many, sometimes complex, issues. The list of issues is then submitted to a representative of the state, and the Committee then asks this representative further questions in the form of a 'constructive dialogue' (Craven 2001: 463). Afterward, the Concluding Observations are drawn up. Their content is comprised of the principal subjects of concern and suggestions and recommendations. Rule 69.1 of the Rules of Procedure allows non-governmental organizations to 'submit to the Committee written statements that might contribute to full and universal recognition and realization of the rights contained in the Covenant' (CESCR 1993: Rule 69.1). As a mechanism for individuals' complaints did not exist till 2013 (see above), written statements or shadow reports by NGOs were of crucial significance. The perception that in the human rights field the governments alone hold the monopoly or reliable information situation in their own country is '[o]ne of the most enduring myths' (Alston 1992: 496).

Craven identified claimed evictions in the Philippines as 'the most remarkable case [. . .] in terms of the potential development of the reporting system' (Craven 2001: 466). In one of its sessions, NGOs informed the Committee about evictions of families in the Philippines. As a consequence, the Committee requested the country respond to this accusation. To get an impression of the workings and language of the Committee, the full paragraph from the Report on the Eighth and Ninth Session of the Committee is quoted:

> The Committee had also received a detailed submission from a coalition of non-governmental organizations relating to alleged violations in the Philippines of the right to adequate housing contained in article 11 (1) of the Covenant. After considering the relevant information, the Committee decided that the information contained in the document was enough to give rise to concern that violations were occurring and that future measures might amount to further violations of the obligations contained in the Covenant. In order to pursue the matter, it asked its Chairperson to address a letter to the Government of the Philippines, requesting it to provide information in response to the document and preferably to do so in the context of a report on articles 10 to 12 to be submitted to the Committee as soon as possible, and in any event prior to its tenth session.

> It noted that, although the Philippines had been a party to the Covenant since 1976, it had yet to submit its initial report relating to articles 10 to 12. In relation to that request, the Committee specifically asked its Chairperson to draw the attention of the Government of the Philippines to General Comment number 4 adopted by the Committee in 1991. (CESCR 1994: para. 374)

After the State Party had submitted the requested report (the Philippines 1994), the Committee reacted in its Concluding Observations:

> The scale of forced evictions and the manner in which they are carried out are of concern to the Committee. The Government itself acknowledges that planned forced evictions may affect up to 200,000 families, and that the Government has identified only 150,000 relocation sites. If these estimates are correct, a very significant number of persons currently threatened with eviction will not receive adequate resettlement. Such a situation would not be compatible with respect for the right to housing. (CESCR_PHL 1995: para. 16)

Craven (2001: 466) emphasized that, in this case, the Committee interpreted its mandate creatively when it was concerned with forestalling a potential future violation without having a specific authorization to do so. Even though some scholars claimed that only a few NGOs have focused on ESC rights (Rosas & Scheinin 2001: 431), the Philippines case demonstrates how the Committee interpreted its role in a creative way: 'In fact, the Committee [. . .] has, in the absence of an official complaint procedure, developed its functions under the reporting procedure to something which is more and more resembling a quasi-judicial complaint procedure' (Rosas & Scheinin 2001: 427).

In order to assist the States Parties to fulfill their reporting obligations, the Committee has published its interpretation of the human rights provision with 22 General Comments between 1989 and 2016. In its first General Comment, the Committee (CESCR 1989: para. 1) emphasized that 'it would be incorrect to assume that reporting is essentially only a procedural matter designed solely to satisfy each state party's formal obligation to report to an appropriate international monitoring body'. On the contrary, the Committee (CESCR 1989: para. 2–9) mentioned seven objectives that the monitoring mechanism should achieve. The states should:

- undertake a comprehensive review of their national legislation, administrative rules and procedures—in particular in the initial report—(first objective);
- monitor the actual situation with respect to each of the ESC rights (second objective);
- provide the basis for the elaboration of their policies (third objective);
- facilitate public scrutiny of government policies with respect to ESC rights and encourage the involvement of various sectors of society in the formulation, implementation and review of the relevant policies (fourth objective);
- provide a basis for an effective evaluation (fifth objective);
- enable themselves to develop a better understanding of the problems and the shortcomings with respect to the progressive realization of ESC rights (sixth objective); and
- facilitate the exchange of information among other states (seventh objective). (CESCR 1989: para. 2–9)

Considering the objectives, Alston (1990: 371) noticed that the strength of the reporting procedure did not primarily lie in the formal exchange between the states and the Committee, but rather in the mobilization of domestic political and other forces to participate in monitoring the government policies. General Comment No. 2 (CESCR 1990a) encompasses the Committee's view with respect to international technical assistance measures by realizing ESC rights. In General Comment No. 2, the Committee suggested

> to include virtually all United Nations organs and agencies involved in any aspect of international development cooperation. It would therefore be appropriate for recommendations in accordance with article 22 to be addressed, inter alia, to the Secretary-General, subsidiary organs of the Council such as the Commission on Human Rights, the Commission on

> Social Development and the Commission on the Status of Women, other bodies such as UNDP, UNICEF and CDP, agencies such as the World Bank and IMF, and any of the other specialized agencies such as ILO, FAO, UNESCO and WHO. (CESCR 1990a: para. 2)

As Article 2 ICESCR was regarded of key relevance for the Covenant, the Committee evaluated the nature of States Parties' obligations under Article 2, para. 1, ICESCR in General Comments No. 3 (CESCR 1991b). The Committee (CESCR 1991b: para. 1) repeated how important Article 2, para. 1 is to a full understanding of the Covenant. With respect to the concept of progressive realization of ESC rights, the Committee stated:

> The concept of progressive realization constitutes a recognition of the fact that full realization of all economic, social and cultural rights will generally not be able to be achieved in a short period of time. In this sense the obligation differs significantly from that contained in article 2 of the International Covenant on Civil and Political Rights which embodies an immediate obligation to respect and ensure all of the relevant rights. Nevertheless, the fact that realization over time, or in other words progressively, is foreseen under the Covenant should not be misinterpreted as depriving the obligation of all meaningful content. (CESCR 1991b: para. 9)

In most of the other General Comments, the Committee either concretized its understandings of specific ESC rights, such as the right to work (CESCR 2006) and the right to social security (CESCR 2008), or the Committee explained ESC rights in the context of special groups, such as the elderly (CESCR 1995). Two General Comments evaluate housing. As the comments are regarded as discourse fragments, the Committee's view on housing will be discussed in detail in Chapter 5.

In its first session in December 1986, the Committee adopted Reporting Guidelines (CESCR 1986) to assist the States Parties in submitting their reports. The guidelines have been modified twice (CESCR 1991b and 2009). They are similar to questionnaires with open-ended questions and therefore regarded as part of the dispositif. Their purpose is to advise States Parties on the form and content of their reports and to ensure that the reports are comprehensive and presented in a uniform manner (CESCR 2009). The guidelines 'show a good understanding of the central issues pertaining to each right, avoid the necessity of asking for basic factual information, and enable the Committee to pinpoint the crucial issues more effectively' (Craven 2002: 65–66).

The contents of the guidelines shall here be presented with a special focus on the reporting obligations related to the right to housing. In the first Reporting Guidelines from 1986, the Committee suggested with respect to Articles 10 to 12 ICESCR that the

> Governments should describe the basic conditions prevailing in their countries as well as the basic programmes and institutions relevant to articles 10 to 12, and that they should stress the developments regarding these conditions since the entry into force of [the] covenant. (CESCR 1986: para. 29)

In paragraph 39, the Committee listed the information it demanded with regard to housing. The Committee asked for principal laws, administrative regulations and statistical information; measures taken to expand the housing construction; measures to solve the problems of housing, water supply and sanitary conditions in rural areas; and measures for protecting tenants (CESCR 1986, para. 39).

In 1991, '[i]n response to the recent introduction of an entirely new reporting cycle and in the light of various inadequacies in the approach reflected in the original guidelines' (CESCR 1991b: 1), the Committee adopted new guidelines. The first version was considered as too old and 'the Committee hoped to ease the reporting burden on States by simplifying the guidelines and providing a consolidated general section to be used in all human rights reporting systems' (Craven 2002: 65). More information was requested regarding the right to an adequate standard of living. The Committee presented a more detailed question catalog. The Committee connected the right to an adequate standard of living with poverty and requested the states give statistical information about a poverty line (CESCR 1991b: 11). The right to housing is discussed in the section about Article 11 of the Covenant. The Committee asked for statistical information about seven groups:

(i) homeless people, (ii) people inadequately housed without access to basic amenities, (iii) people living in illegal settlements or housing, (iv) evicted persons and persons currently lacking protection against arbitrary eviction, (v) people whose housing expenses are above any government-set limit of affordability, and (vi) persons on waiting lists for accommodation. Additionally, the Committee asked for laws affecting the right to housing in 13 extra bullet points, such as (ii) homeless persons acts, (iii) land planning and expropriation legislation, (iv) security of tenure and protection from eviction, (ix) legislation to restrict speculation on housing or property (particularly when the speculation has negative effects on housing rights), (x) legislative measures conferring legal title to those living in the illegal sector, or (xi) environmental planning and health in housing and human settlements. Furthermore, the Committee suggested that States Parties shall report about different measures such as building housing, releasing land, financial measures, or measures to encourage the development of small and intermediate centers, especially at the rural level. In 2009, the Committee decided to replace the second version of the Reporting Guidelines

> to take into account the harmonized guidelines on reporting under the international human rights treaties, as well as the evolving practice of the Committee in relation to the application of the Covenant, as reflected in its Concluding Observations, general comments and statements. (CESCR 2009: para. 3)

The third version of the Reporting Guidelines was shorter than the former versions. The requested information for the right to an adequate standard of living again contained statistical information about a poverty line or other mechanisms to measure the incidence and depth of poverty (para. 42) and—this was new—a national action plan or strategy (para. 43a) and targeted policies and programs (para. 43b) to combat poverty. Moreover, information concerning the right to water was requested in a special single paragraph (para. 44). The requested information about the right to housing was listed in paragraphs 50–54. Paragraph 50 should here be regarded with particular attention:

> Indicate whether a national survey on homelessness **and** inadequate housing has been undertaken, as well as its findings, in particular the number of individuals and families who are homeless **or** inadequately housed and without access to basic infrastructures and services such as water, heating, waste disposal, sanitation, and electricity, as well as the number of persons living in over-crowded or structurally unsafe housing. (CESCR 2009: para. 50, author's emphasis)

Special consideration should be given to the marked words in the paragraph. The Committee talked about homelessness and inadequate housing. Moreover, the Committee mentioned people who are homeless on the one hand, and people who are inadequately housed and have no access to basic infrastructure services on the other hand. It seems that the Committee distinguished between two main categories relevant for the human right to housing: inadequate housing in general and homelessness as a sort of no housing.

Paragraphs 51 and 52 reflect six of the seven aspects that also came up in General Comment No. 4 (Table 5.1, p. 134). The Committee requested information concerning measures about (a) the affordability of housing, (b) the impact of social housing, and (c) the accessibility and habitability for persons with special housing needs such as persons with disabilities, older people, and families with children. Paragraph 52 deals with the aspect of location, and the States Parties were asked to report about 'legislative and other measures in place to ensure that housing is not built on polluted sites or in immediate proximity to pollution sources that threaten the health of inhabitants' (para. 52). Paragraphs 53 and 54 focus more on forced eviction. Only the aspect of cultural adequacy was not mentioned directly.

The differences between the three guidelines should not only be explained with a differentiating consideration of the content and scope of the right to housing, but also with a different consideration of the role of the guidelines. The three guidelines were, in general, different with regard to their focal points. In the first version, the focal point was on measures; the 1991 guidelines were detailed and asked for a lot of statistical information, laws, and measures, while the 2009 guidelines did not focus as much on statistics and were considerably shorter than the first two versions.

This section has emphasized the major role of the Committee in the ICESCR discourse. It has shown that the Committee is, in comparison with the states, the more powerful speaker. Or, as Craven stressed:

> If the Committee is said to have any role in 'implementation' of economic, social and cultural rights, it is primary insofar as it has undertaken the task of developing an authoritative understanding of the nature and content of those rights and of promoting that understanding internationally. (Craven 2001: 472)

2.4 ICESCR Member States

The underlying database for the present research consists of 386 full States Parties reports of 132 states from all UN regions (Database FLOOR A, SPR) plus the Concluding Observations for 131 States Parties (Database FLOOR A, CO) (Table 2.2). This also includes states that do not exist anymore such as Czechoslovakia, the German Democratic Republic, and Yugoslavia as well as states that changed their names and their territorial borders. The Russian Federation and Russia are counted as one state, for instance. This is also true for former West Germany and the Federal Republic of Germany as it exists now. The official name of the Federal Republic of Germany did not change in 1990, but only included the former five states from the Democratic Republic of Germany. Until 1986, the states submitted their reports in three sections. In the first section, the states reported on rights covered by articles 6–9. In the second section, they reported on rights covered by articles 10–12 and in the third section on rights covered by articles 13–15. These reports are summarized into one report, even if the country only finished one or two of the sections before switching to the system of submitting full reports. When a State Party combined two (such as the Netherlands) or even three

Table 2.2 States Parties reports and Concluding Observations

UN region	No. of submitting states	No. of submitted full reports	No. of states that received Concluding Observations
Africa	28	66	31[1]
Americas	24	79	24[2]
Asia	32[3]	85	31[4]
Europe	45	148	42[5]
Oceania	3	8	3
Sum	**132**	**386**	**131**

Source: Database FLOOR A, SRP & CO. The list of the UN regions is available at http://unstats.un.org/unsd/methods/m49/m49regin.htm.

[1] There are no Concluding Observations for Uganda, but there are Concluding Observations for Congo, Equatorial Guinea, Guinea, and Mali. These states did not submit a report yet.

[2] There are no Concluding Observations for Barbados, but there are Concluding Observations for Saint Vincent and the Grenadines. This state did not submit a report yet.

[3] Including Nepal.

[4] There are no Concluding Observations for Thailand.

[5] There are no Concluding Observations for Czechoslovakia, German Democratic Republic, and Yugoslavia.

reports (such as Peru), this counted as two (or three) reports. I did not analyze States Parties reports for the Dependent Territories of the United Kingdom, the Netherlands' Antilles, and Macao, which is a territory under the Portuguese administration. Greenland is an autonomous country within the Kingdom of Denmark. Denmark included sections concerning Greenland in its reports. Therefore, in the comparative discussion (including the diffusion maps), Greenland was treated as part of Denmark.

The differentiation between UN regions is based on the compositions of macro-geographical (continental) regions. In the comparative discussion, I will point out geographical differences and overlaps, but will not differentiate between other groups such as Organisation for Economic Co-operation and Development (OECD) countries and non-OECD countries. Diffusion Map 2.1 illustrates the ICESCR member states and shows in which decades they submitted their first State Party report.

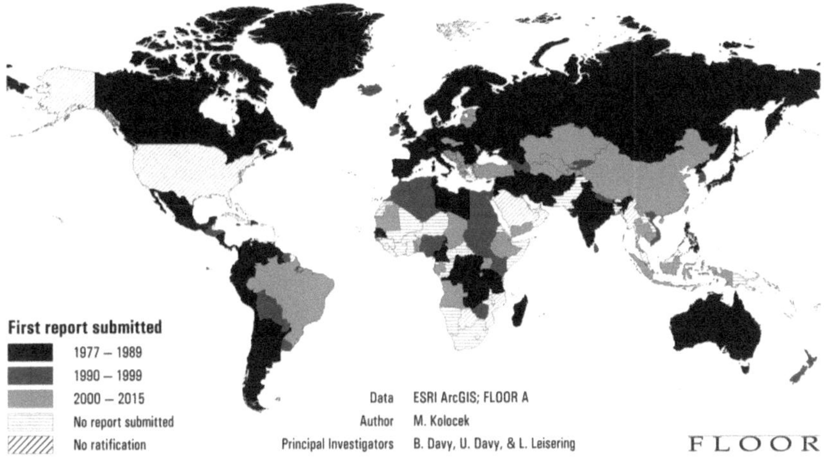

Diffusion Map 2.1 Year of submission of the first ICESCR State Party report

Moreover, it shows the states that have not ratified the Covenant and the states that ratified the Covenant but have not submitted a report yet. This includes in particular African states. The UN Region of the Americas includes states from Latin America and the Caribbean, plus Canada.

By analyzing the reports filed under ICESCR, U. Davy (2014: 243) found that 40 out of 160 member states have not submitted a report yet, and the submitted reports have an average delay of 42 months. Moreover, 'the substantial parts of the reports are evasive, inconclusive, or lack data' (U. Davy 2014: 243). Craven (2001: 464) came to a similar result and counted 97 overdue reports from 88 States Parties and 17 States that have failed to submit a single report in 10 years (till May 1996).

There is no significant correlation between the wealth of a state and its member status of the Covenant. From the group of 25 states with the highest Human Development Index (HDI) (UNDP 2014: 160), the United States is the only one that is not a member of the Covenant. From the group of the 25 states with the lowest HDI (UNDP 2014:

162–163), the only non-member state is Mozambique. However, there seems to be a correlation between the HDI and the submitting discipline. While all 24 ICESCR members with the highest HDI have submitted at least one report, 11 of the 24 ICESCR members with the lowest HDI had not submitted a report by 2015. Incidentally, all 11 of those countries are in Africa. Craven (2002: 58) proposed several main reasons for the weak submission rates among developing countries. In these countries, internal organization expertise and other personal and economic resources were often missing, and meanwhile, the original set of guidelines was complicated, general, and difficult to apply.

The scholarly literature explaining why the United States signed, but never ratified the ICESCR, is scarce. One explanation states that the US governments regarded social rights as merely desirable goals or as 'so-called rights' (Alston 1990: 367). Piccard claimed that '[i]f the United States ratifies no other human rights treaty this century, it ought to ratify the ICESCR' (Piccard 2010: 233). By trying to explain the non-participation of the country, Piccard (2010: 243) acknowledged that a ratification that would be followed up by an enactment of implementing legislation would dramatically alter the civil, political, economic, social, and cultural landscape in the United States. Moreover, the country would then 'open itself to scrutiny by the International community via the Committee on Economic, Social and Cultural Rights' (Piccard 2010: 249). The American skepticism may also result from the ideological conflict between East and West during the drafting process of the Covenant. Alston accused many Americans of suspecting that the Covenant was not an international treaty to promote the satisfaction of material needs, but rather a 'Covenant on Uneconomic, Socialist and Collective Rights' (Alston 1990: 366). The author specifically addressed the social rights to an adequate standard of living (Article 11 ICESCR), health (Article 12 ICESCR), and education (Articles 13 and 14 ICESCR) as problematic from the US perspective and questioned:

> Is the United States prepared to commit itself to the general proposition that there is indeed a human right to each of these social goods or, put differently, to the satisfaction of each of these basic human needs? And

> even if it is, is it prepared to accept the specific level of obligation in that regard provided for by the Covenant? (Alston 1990: 369)

It thus seems that the United States did not neglect the importance of ESC rights; it rather neglected the proposition that the state ought to be the central actor in guaranteeing these rights.

2.5 Human Rights between the Spoken Word and the Spatial Reality

All in all, the ICESCR is recognized in most countries worldwide, and its significance is increasing. As indicated, nowadays, a vast amount of data exists, produced by many supranational actors—UN sponsored actors, international non-governmental organizations (INGOs), and complex bodies—to clarify the content of the human right to housing and to investigate its implementation in different countries worldwide. When it comes to the implementation of human rights, the states are always the first actors mentioned. The states' consideration of the human right to housing is of essential significance with respect to its implementation. The monitoring mechanism under ICESCR shines a spotlight on the states' way of articulating what their housing problems are. At present, more than one hundred States Parties from all UN regions have submitted States Parties reports, so this database allows a global comparative investigation of the consideration of the human right to housing that is hardly achievable with any other data source.

Until today, '[the] meaning of human rights is not fixed once and for all by the ideas of lawmakers. Human rights are meant to be "living instruments" [. . .]. The content of human rights is constantly negotiated anew among relevant actors' (U. Davy 2013: S23). The contents and the meaning of the human right to housing have changed during the past decades. As the first States Parties reports were submitted in the mid-1970s, the data source does allow not only a comparative analysis between the member states, but also an in-depth investigation of the changing considerations. The analysis will show that, during nearly

40 years, ingredients of social citizenship and even global social citizenship have gradually spread into the states' way of talking about the human right to housing.

The concept of discourse analysis highlights the potential of the spoken word as an issue for analysis, and, at the same time, does not neglect the policy level. By emphasizing that 'in the practical use of signs, meanings of phenomena are socially constructed and these phenomena are thereby constituted in their social reality' (Keller 2013: 3; see above), discourse analysis helps understand the relationship between the spoken word and the spatial reality.

The tools and terms offered by discourse analysis furthermore separate the complexity of speaking and non-speaking actors and documents in a manner that makes the analysis transparent for other readers. Moreover, discourse analysis offers an interpretative repertoire that helps to reconstruct the reports in the face of theories concerning land, social citizenship, and social policy, even on a global level.

3

Discourses on Housing

This chapter builds a bridge between the right to housing and social policy. Section 3.1 examines housing from the perspective of the law. I discuss the states' and others' obligations and point out international and national court cases. These cases draw attention to the interdependence of the right to housing and other rights and needs, especially the right to property. Section 3.2 discusses housing as a subject of several studies under the umbrella of discourse analysis. Then, I show that some time Western scholars' housing research focused on the question of owning versus renting. I then consider the tenure forms of homeownership versus renting in the context of housing satisfaction (Section 3.3). The tenure question influences housing satisfaction: Homeowners are usually more satisfied with their housing situation than tenants. I discuss how the housing satisfaction in countries with a low ownership rate is nevertheless higher than in countries with a high ownership rate. By presenting two main forms of inadequate housing, that is, homelessness and Spaces of Inadequate Housing (Section 3.4), I then show that inadequate housing often stands outside of formal property or renting markets.

M. Kolocek, *The Human Right to Housing in the Face of Land Policy and Social Citizenship*, DOI 10.1007/978-3-319-53489-3_3

3.1 Housing and the Law

The right to housing has found its way into the ICESCR (Section 2.2) and other international treaties (Article 17 ICCPR), especially for specific groups: the Convention on the Elimination of All Forms of Discrimination against Women (Article 14, para. 2 h, CEDAW), the Convention on the Rights of the Child (Article 27, para. 3, CRC), the Convention on the Rights of Persons with Disabilities (Articles 9 and 28 CRPD), the Convention Relating to the Status of Refugees (Article 21 1951 Convention), the Convention on the Elimination of All Forms of Racial Discrimination (Article 5, para. e [iii], CERD), and the Convention on the Protection of the Rights of All Migrant Workers and Members of Their Families (Article 43, para. 1d, MWC) (see also Leckie 2001; UN-Habitat 2009a; UN-Habitat & OHCHR 2002). On a European level, the right to housing has found its way into the Charter of Fundamental Rights of the European Union (Article 34, para. 3, CFR), the European Social Charter (Article 31 ESC), and other conventions for specific groups such as migrant workers or children (see, for a more detailed list UN-Habitat 2009a: 12). As a term, the right to housing is not mentioned in the American Convention on Human Rights (ACHR), the African Charter on Human and Peoples' Rights (ACHPR), or the Arab Charter of Human Rights (Arab CHR). Housing rights, however, can be inferred from other rights, such as the right to a home (Article 8 European Convention for the Protection of Human Rights and Fundamental Freedoms [also European Convention on Human Rights or ECHR]) or the right to private property (Article 21 ACHR; Article 25 Arab CHR; Article 14 ACHPR; Article 1 Protocol 1 ECHR). In addition, some ILO conventions and resolutions adopted by the United Nations mention housing or property rights (see UN-Habitat & OHCHR 2002: 4–8).

On national levels, housing was incorporated into the constitutions of more than 40% of the countries in the world (UN-Habitat & OHCHR 2002: 36–37). Oren et al. (2014) observed the wording of the right to housing in 205 national constitutions. They distinguished between three major forms by which countries incorporated housing rights into their constitutions: (1) housing as a direct right, that is, either the right to housing as a stand-alone article or housing listed among other social rights; (2) an embedded form of housing rights, namely rights that refer to housing as a component of other rights; rights of specific groups such as workers, children, or the disabled; welfare responsibilities of the state; and more than twenty other listed examples; and (3) indirect or implied rights. This is an indirect form of protection of housing rights which does not include an explicit mention of housing. The authors regarded property rights as the most prominent example. They stated that the embedded form of constitutional housing rights is the most widespread (Oren et al. 2014: 146). By taking such an approach it seems impossible to imagine a constitution that completely ignores rights that at least indirectly relate to housing.

There is less literature on housing jurisprudence in the UN regions of Africa, Asia, the Americas, and Oceania than in Europe and North America. The Centre of Housing Rights and Evictions (COHRE 2002) summarized housing and property relevant cases considered by the Inter-American Commission on Human Rights and the Inter-American Court of Human Rights. Most of the cases before the Inter-American Commission on Human Rights consider the demolition of property during armed conflicts, unlawful imprisonment, or indigenous land rights (COHRE 2002: 21–30). COHRE (2002: 31) stated that the Inter-American Court on Human Rights has considered relatively few housing relevant cases. The most quoted housing cases from the African continent are from South Africa (van der Walt 2009). This is because the African Court on Human and Peoples' Rights is a young institution that delivered its first judgment in 2009 (FIDH 2010: 15). For Asia, there is no regional counterpart to the European Court of Human Rights (ECtHR), the African Court on Human and Peoples' Rights, or the Inter-American Court of Human Rights. This is also true for Oceania.

Considering the aforementioned international treaties, the right to housing is above all an obligation, or duty, of the state. But what does this mean and how far does this obligation go? With respect to the rights in the ICESCR, two documents give answers to these questions. During June 2–6, 1986, a group of distinguished experts in international law met in Maastricht 'to consider the nature and scope of the obligations of

States parties to the International Covenant on Economic, Social and Cultural Rights, the consideration of States parties' reports by the newly constituted Committee on Economic, Social and Cultural Rights, and international cooperation under Part IV of the Covenant' (Limburg Principles 1987: Introduction). The 29 participants, including four members of the newly constituted Committee, came from countries all over the world (except from Asia), and represented different global actors: United Nation Centre for Human Rights, ILO, UNESCO, WHO, the Commonwealth Secretariat, and sponsoring organizations (Limburg Principles 1987: Introduction). The expert group considered the nature and the scope of the States Parties' obligations regarding Articles 2 to 5 and Article 8 of the Covenant and the reporting process including the role of the Committee and its relations between specialized agencies and other international organs. Considering the scope of the ICESCR member states' obligations, this document is one of the first that interpreted prominent excerpts of the Covenant such as 'to achieve progressively the full realization of the rights' or 'to the maximum of its available resources' from Article 2 of the ICESCR (Limburg Principles 1987: para. 21–28). The Limburg Principles have been issued as an official United Nations document after the initiative of the Dutch government (Flinterman 1997: 244).

Another attempt to enhance the legal status of ESC rights is the publication of the Maastricht Guidelines in 1997. The Maastricht Guidelines were adopted by a group of more than 30 experts on the tenth anniversary of the Limburg Principles. They were

> designed to be of use to all who are concerned with understanding and determining violations of economic, social and cultural rights and in providing remedies thereto, in particular monitoring and adjudicating bodies at the national, regional and international levels. (Maastricht Guidelines 1997: Introduction)

This workshop concentrated on the relevance 'of a "violations approach" in order to strengthen the monitoring system' (Flinterman 1997: 245). The Limburg Principles and the Maastricht Guidelines are key documents to supplement the understanding of the Covenant because they helped the states, the Committee, the NGOs, and many other actors to specify how ESC rights should be understood. The States Parties' obligations were separated into the obligation to respect, protect, and fulfill ESC rights:

> Failure to perform any one of these three obligations constitutes a violation of [...] rights. The obligation to **respect** requires States to refrain from interfering with the enjoyment of economic, social and cultural rights. Thus, the right to housing is violated if the State engages in arbitrary forced evictions. The obligation to **protect** requires States to prevent violations of such rights by third parties. Thus, the failure to ensure that private employers comply with basic labour standards may amount to a violation of the right to work or the right to just and favourable conditions of work. The obligation to **fulfil** requires States to take appropriate legislative, administrative, budgetary, judicial and other measures towards the full realization of such rights. (Maastricht Guidelines 1997: para 6., author's emphasis)

The respect-protect-fulfill typology has also been transferred to the right to housing. The Commission on Human Settlements (1999: para. 77) tried to transfer the obligations to possible housing actions and added the obligation to promote. In 2009, UN-Habitat rejected the category to promote. Instead, UN-Habitat substantiated its understanding of the three obligations. The obligation to respect required the states to refrain from interfering directly or indirectly with the enjoyment of the right to housing, for instance through refraining from carrying out forced evictions or denying the security of tenure for particular groups (UN-Habitat 2009a: 33). Leckie (2001: 156) stated that the obligation to respect shall also include the obligation to respect peoples' rights to build their own dwellings; even if the self-built housing is outside legal regulations. Dankwa et al. (1998) commented on each of the Maastricht Guidelines. By considering the first and second aspect, they suggested:

> Respect for and protection of economic, social and cultural rights will be made more meaningful if victims of violations of these rights are not punished for the state in which they find themselves, a condition for which others bear responsibility. To illustrate, a refugee or asylum seeker

should not be made to bear criminal liability for being homeless. (Dankwa et al. 1998: 726)

The obligation to protect was understood as an obligation to prevent third parties, for instance, private actors, from interfering with the right to housing. The states should ensure that private actors comply with human rights standards related to the right to housing (UN-Habitat 2009a: 33). The obligation to fulfill is 'the most positive in nature' (Leckie 2001: 157) and was interpreted as the obligation to adopt housing policies or a national housing plan that included lots of single measures, inter alia the provision of physical infrastructure for housing (UN-Habitat 2009a: 33–34). The fulfill aspect, so asserted Dankwa et al. (1998: 714) 'relates closely to the duty of states to devote the maximum of available resources towards the progressive realization of the rights established under Article 2(1) of the ICESCR.'

The Committee on Economic, Social and Cultural Rights is not the only global actor that monitors the implementation of ESC rights on a global level. Considering the emergence of complex international actors that deal with housing, one could state that Deacon's (2007) hypothesis about the globalization of social policy (4.1) has been proven correct in the case of housing. During the last few decades, many different international actors emerged that, inter alia, monitor the implementation of the human right to housing. They all view inadequate housing in a supranational context. Mostly, their publications summarize case studies from different states and regions of the world. Three such global actors shall be presented in a few words: UN-Habitat, the Special Rapporteur on the right to adequate housing, and the European Federation of National Organisations Working with the Homeless (FEANTSA).

FEANTSA is a cooperation of different European and non-European states, researchers, and national organizations that work with and on behalf of homeless people. Its basic aims are to provide information and to analyze homelessness, to monitor national policies and give recommendations (Avramov 1995: 1).

UN-Habitat was established after the Vancouver conference on human settlements (1976) by a UN resolution (UN 1977) and is the

United Nations agency for human settlement. It is mandated by the UN General Assembly to promote socially and environmentally sustainable towns and cities with the goal of providing adequate shelter. In 2002, the United Nations General Assembly decided to elevate UN-Habitat into a fully-fledged program of the United Nations (UN 2002). Since 1986, UN-Habitat has produced Global Reports on Human Settlements (e.g., UN-Habitat 2003 and 2009b) and many other documents related to housing, often in cooperation with other agencies.

The Special Rapporteur on the right to adequate housing was appointed in April 2000 by the Commission on Human Rights for a period of 3 years to focus on adequate housing as a component of the right to an adequate standard of living (CESCR 2002b: para. 7c). The Special Rapporteur is, among other things, requested to

- report on the status, throughout the world, of the realization of the rights that are relevant to the mandate;
- promote, as appropriate, cooperation among and assistance to governments in their efforts to secure these rights;
- apply a gender perspective of his work;
- develop a regular dialogue with involved actors like the governments, Habitat, NGOs, or financial institutions; and
- to submit to the Commission an annual report covering the activities related to this mandate (CESCR 2002b: para. 7d).

The Special Rapporteur's intention is 'to promote greater realization and operationalization of the right to adequate housing through a constructive approach, by closing the gap between legal recognition and practice and seeking solutions to the grave housing and living conditions found globally' (ECOSOC 2001: 3). As the present analysis shows, the Committee on Economic, Social and Cultural Rights used the Special Rapporteurs' reports as source in its Concluding Observations.

So far, the focus was primarily on the states' duties. Human rights are, indeed, more than just claims of individuals against the state.

> Human rights are *rights*; they are not merely aspirations, or assertions of the good. To call them rights is not to assert, merely, that it is 'right' that the individual shall enjoy these goods; or even, merely that it is the duty of society to respect the immunity or provide the benefits. To call them 'rights' implies that they are claims 'as of right,' not by appeal to grace, or charity, or brotherhood, or love; they need not to be earned or deserved. The idea of rights implies entitlement on the part of the holder in some order under applicable norm; the idea of human rights implies entitlement in a moral order under a moral law, to be translated into and confirmed as legal entitlement in the legal order of political society. When a society recognizes that a person has a right, it affirms, legitimates, and justifies that entitlement, and incorporates and establishes it in the society's system of values, giving it important weight in competition with other societal values. (Henkin 1990: 3, emphasis in the original)

Henkin used the term society for good reasons. To implement housing (or other human) rights usually means to balance the rights and duties of several parties involved. One of my main arguments is that private property rights are sometimes the cause and sometimes the solution for inadequate housing. The argument is not new, but still worth evaluating. Examining housing and property through the lens of law means to figure out the rights and duties of those who own the land (the self-using private property owner, the landlord, the state, the municipality, or the developer) and the rights and duties of those who do not (the homeless person, the squatter, or the tenant).

Many court cases of housing rights violations concern evictions or expropriations. When it comes to expropriations or evictions (or both), the courts consider questions such as: Who is the owner of the land? What is the purpose for the expropriation/eviction and what is the designated future land use? If there is economic development, who profits? Is there any compensation, and if yes, how much? What happens to the evicted/expropriated persons? Van der Walt (2011) compared jurisdictions on expropriation and eviction from Germany, South Africa, the United Kingdom, the United States, and the European

Court of Human Rights. He evaluated the obligations of the state and the local authorities regarding the provision of housing. Van der Walt (2011: 98–99) emphasized that the national housing legislations must always be read in the country's historical and constitutional context. Concerning the similarities and differences between the jurisprudence in South Africa and the other countries, van der Walt (2011: 99) stated that other countries did not have the same historical baggage to deal with (see also Strauss & Liebenberg 2014). The author discussed the meaning of home and concluded:

> Protection of the home interest cannot always trump the power to expropriate private property for a public purpose, but when homes and communities are destroyed for a public purpose the expropriation should actually serve a real and serious public purpose, and when the public purpose relates to economic development the affected residents and community should benefit from it. (Van der Walt 2011: 100)

The idea of the right to a home has become more and more prominent in discourses on the right to housing (Fox O'Mahony & Sweeney 2011). This is also true for the right to the city which has its roots in the famous essay by Henry Lefebvre (1968) and has returned to the agenda in the last few years (Harvey 2013; Kothari & Chaudrhy 2012: 41). Kenna demanded a meaningful legal concept of home to enhance housing rights approaches:

> Perhaps it is [. . .] time for a fresh consideration of the meaning of rights and the historical and ideological alignment of housing rights with property rights. The distinction can be seen more clearly when housing rights begin to incorporate the concept of home rather than property. Indeed, the concept of 'home' itself is much wider than a legal concept, and involves important subjective, cultural, emotional, social status, and social relational issues. It involves more than a permanent or temporary dwelling, but includes the human dimension of living and having relationships. (Kenna 2008: 468)

The enforcement power of the right to housing in the international treaties depends on each treaty's monitoring mechanism. The European Social Charta (ESC) and its successor, the Revised European Social Charta (RESC) supplement the European Convention on Human Rights (ECHR) in the field of social and economic rights. They are the regional counterpart to the ICESCR (Harris 2009: 3–4) because, inter alia, the monitoring mechanism is similar. The member states have to submit reports to the European Committee on Social Rights (ECSR), established under Article 25 ESC. In addition, states that ratified the Additional Protocol to the European Social Charter Providing for a System of Collective Complaints (AP-ESC) recognize the right of NGOs (and other organizations) to submit complaints alleging unsatisfactory applications of the Charter (Article 1 AP-ESC). The European Federation of National Organizations working with the Homeless (FEANTSA) submitted such a complaint (ECSR 2007: FEANTSA v. France) and accused France of not ensuring an effective right to housing as derived from Article 31 RESC:

> In particular, [FEANTSA] considers that the measures in place in France to reduce the number of homeless people are insufficient, that the construction of social housing is also insufficient, that a significant number of households live in poor housing conditions, notably with regard to sanitation and overcrowding, and argues that the implementation of legislation on the prevention of evictions is dysfunctional. FEANTSA also alleges that the system for allocating social housing and the associated remedies do not function properly and that there is discrimination in access to housing with regard to immigrants. (ECSR 2007: para. 17)

The French government maintained that Article 31 RESC only requires states 'to take measures, not to achieve results' (ECSR 2007: para. 18), an argument that sounds familiar to those who analyzed the 'progressive realization' wording from Article 2, para. 2, ICESCR in detail (Saul et al. 2014: 133–172). The ECSR considered the French housing legislation and other relevant legislation, that is, legislation on tenancy, anti-exclusion, family rights, building, planning, local responsibilities, and more. The ECSR agreed with the French government's argument about the government's responsibility to take measures (instead of achieving goals), but stated that 'rights recognized in the Social Charter must take a practical and effective, rather than purely theoretical, form' (ECSR 2007: para. 55). FEANTSA presented statistics on persons living in dwellings with no basic amenities and persons living in conditions of overcrowding (para. 68–75), evictions and homelessness (para. 94–99), social housing units (para. 111), and the housing situation of migrant households and travelers (para. 148–154). After considering the French government's point of view, the ECSR held that there was a violation of Article 31 RESC with respect to all points FEANTSA has claimed:

- insufficient progress in the eradication of substandard housing and lack of proper amenities of a large number of households;
- unsatisfactory implementation of the legislation on the prevention of evictions and the lack of measures to provide rehousing solutions for evicted families;
- insufficient measures currently in place to reduce the number of homeless persons;
- insufficient supply of social housing accessible to low-income groups;
- malfunctioning of the social housing allocation system; and
- deficient implementation of legislation on stopping places for travelers. (ECSR 2007: Conclusion)

The complaint by FEANTSA draws attention to the states' and other responsibilities. When it comes to the right to housing, governmental actors should do more than just implement housing laws. Although they cannot be blamed for not achieving certain goals, their responsibility goes beyond the theoretical recognition of rights. As the ECSR stated, the rights in the RESC must take a practical and effective form. In the language of the respect-protect-fulfill triangle, one could read the ECSRs decision as an emphasis on the states' duty to respect, protect, and (!) fulfill the right to housing. In addition, this example shows the role (and power) of NGOs in the implementation of the right to housing. A lot of NGOs work in the field by directly supporting people affected by inadequate housing in different ways. FEANTSA here acted 'on a discourse level' to claim and prove a violation of housing rights in France. However, France is one of only a few European countries that have even signed the AP-ESC (Harris 2009: 4) which is a precondition for such a complaint.

For Europe, Kenna and Jordan (2014: 116) found a growing significance of housing rights in the context of homelessness, poor-quality housing, access and affordability, and social segregation. Housing has also been negotiated before the European Court of Human Rights (ECtHR) in Strasbourg, although the right to housing is not directly mentioned in the ECHR. Most of the housing relevant cases affect Article 8 ECHR (the right to a home) or Article 1 Protocol 1 ECHR (the right to private property). The cases arose from armed conflicts (ECtHR 2014: Cyprus v. Turkey), temporary expropriation (ECtHR 1982: Sporrong and Lönnroth v. Sweden), compensation claims after the land reforms in former Eastern Germany (ECtHR 2005: Jahn and others v. Germany), or the balancing of landlords' duties and tenants' rights (ECtHR 1989: Mellacher and others v. Austria; ECtHR 2006: Hutten-Czapska v. Poland) (see also Kenna 2008; Leckie 2001; Ploeger & Groetelaers 2007; van der Walt 2009 and 2011).

I shall present two cases that went before the European Court of Human Rights in more detail because they demonstrate a remarkable view of illegal housing: Öneryıldız vs. Turkey (ECtHR 2004) and Winterstein and others v. France (ECtHR 2013). After a methane explosion on April 28, 1993, in a disused garbage tip in an informal neighborhood (gececondu) in Istanbul, the applicant Öneryıldız demanded compensation for the death of nine of his relatives and the destruction of his property. The applicant's damaged hut had been built in a slum quarter of Kazım Karabekir in Ümraniye without any authorization. The methane explosion happened due to the area being on land surrounding a rubbish tip used by four councils since the 1970s. Before the Istanbul Criminal Court, two mayors were convicted because no measures were taken to prevent this explosion. Before the ECtHR, Mr. Öneryıldız claimed a violation of, inter alia, Article 2

ECHR (right to live), Article 8 ECHR (right to home), and Article 1 Protocol 1 ECHR (protection of property). The Turkish Government argued that the applicant had been acting illegally when settling by the rubbish tip. The European Court of Human Rights, however, did not accept this argument and stated:

> The information and documents in the Court's possession show that, since 1960, when inhabitants of underprivileged areas started migrating in their masses to the richer regions, Turkey has been confronted with the problem of slums, consisting in most cases of permanent structures to which further floors were soon added. It would appear that currently more than one-third of the population live in such dwellings. Researchers who have looked into the problem maintain that these built-up areas have not sprung up merely as a result of deficiencies in urban planning or shortcomings on the part of the municipal police. They point to the existence of more than eighteen amnesty laws which have been passed over the years in order to regularise the slum areas and, they believe, satisfy potential voters living in these rudimentary dwellings.
>
> [. . .] The Court concludes from these legal considerations that, in spite of the statutory prohibitions in the field of town planning, the State's consistent policy on slum areas encouraged the integration of such areas into the urban environment and hence acknowledged their existence and the way of life of the citizens who had gradually caused them to build up since 1960, whether of their own free will or simply as a result of that policy. Seeing that this policy effectively established an amnesty for breaches of town-planning regulations, including the unlawful occupation of public property, it must have created uncertainty as to the extent of the discretion enjoyed by the administrative authorities responsible for applying the measures prescribed by law, which could not therefore have been regarded as foreseeable by the public. (ECtHR 2004: para. 53 and 104)

Taking Article 1 Protocol 1 ECHR into consideration, the Court had to decide whether there was possession or not. This question is interesting because, although the dwelling was built illegally on public land, it was tolerated by the state for a long time. The Court distinguished between the dwelling built by the applicant and the land below the dwelling which formally belonged to the state. Taking the land into account, the

Court declined a status of it being the applicant's property. Considering the proprietary interest in the applicant's dwelling, the Court recognized 'a substantive interest and hence a "possession"' (para. 129) deriving from Article 1 Protocol 1 ECHR, even if it was built illegally. Ploeger and Groetelaers (2007: 1429) commented on the decision and hypothetically asked whether such an argumentation opened the way for all builders of illegal dwellings to argue that they have possession protected by Article 1 Protocol 1 ECHR if authorities have tolerated their illegal use of land. B. Davy and Pellissery (2013: S69) found it surprising that the applicant won once he framed his claim in terms of human rights; this shows, so they asserted, that human rights fully apply to informal dwellers. The case demonstrates that property, as a bundle of rights, to some extent includes non-formal rights, particularly when connected to housing.

In the next case, Winterstein and others v. France (ECtHR 2013), the applicants were travelers (some of them regarded themselves as 'Gypsies') who were evicted from a plot of land where they had been living for somewhere between five and 30 years. The applicants alleged a violation of Article 8 ECHR. They claimed that they had been living on the land for a long time and that under French law this gave them possession. Ironically, the French Court used the same argument—the length of time they had been occupying the land—to call their status as travelers into question. The case is interesting for several reasons. First, the European Court of Human Rights had to describe its view on the concept of home:

> [T]he concept of 'home' within the meaning of Article 8 is not limited to premises which are lawfully occupied or which have been lawfully established. It is an autonomous concept which does not depend on classification under domestic law. Whether or not a particular premise constitutes a 'home' which attracts the protection of Article 8 will depend on the factual circumstances, namely, the existence of sufficient and continuous links with a specific place. (ECtHR 2013: para. 69)

This concept, stressed by the Court, is interrelated with the applicants' right to respect their private and family life, and their identity.

> The Court observes that the present case also brings into play, in addition to the right to respect for one's home, the applicants' right to respect for their private and family life, as the Government implicitly recognised. It reiterates that the occupation of a caravan is an integral part of the identity of travellers, even where they no longer live a wholly nomadic existence, and that measures affecting the stationing of caravans affect their ability to maintain their identity and to lead a private and family life in accordance with that tradition. (ECtHR 2013: para. 70)

The Court finally decided that there was a violation of Article 8 ECHR. It explicitly stated that it was not in dispute that the applicants had been living on the land for many years. The Court's main argument was that the domestic courts ordered the eviction without having analyzed the proportionality of this measure: 'Once [the domestic courts] had found that the occupation did not comply with the land-use plan, they gave that aspect paramount importance, without weighing it up in any way against the applicants' arguments' (ECtHR 2013: para. 84). This argumentation is different from the Öneryıldız case, in which the numerous years of tolerating informal housing structures have justified a right to property deriving from Article 1 Protocol 1 ECHR. As the applicants did not claim a violation of their right to property, this question was not raised in the Winterstein case.

One of the most quoted and highly debated cases from the United States is Kelo v. City of New London (US Supreme Court 2005). The US Supreme Court declared expropriation (taking) of land for the sake of an economic development plan lawful. The area was in economic decline. The Court stated, inter alia, that the plan served a public purpose and that the expropriation satisfied the public use requirement of the US takings clause. The case is not only fascinating due to the decision as such, but also for what happened afterward. No redevelopment has ever taken place, developers missed their final deadline for securing financing for building, and Susette Kelo's house was saved and moved to a plot not far from the former plot (Van der Walt 2011: 64–65). Due to the Kelo case and its consequences, most of the states in the country enacted reform legislation, often inhibiting expropriations for the sake of economic development (Van der Walt 2011: 66). The Kelo case was not a case on housing rights in the first place. The main issue at stake was about public use requirements. The case, however, illustrates the complex relationship between economic interests, public use requirements, and housing and property rights. Of course, it would be too simple to use failed planning as a main argument to prove that the Court was wrong. Perhaps, though, the Kelo case demonstrates how difficult it is to foresee future developments.

When balancing tenants' rights and landlords' duties, Germany is an interesting country because Germany has a low rate of home ownership which is usually explained as renting being an attractive option for German households (3.3). This is true for the costs of renting (Kurz 2004: 22) and for the tenants' rights. The Federal Constitutional Court (1993) even regarded tenancy rights as property within Article 14 of the German Basic Law. A few years ago, the Local Court Düsseldorf (Düsseldorfer Amtsgericht) had to decide a question that is usually debated in other discourse areas, namely whether men can stand while they are urinating (Düsseldorfer Amtsgericht 2015). What happened? A landlord had been accused by his former tenant of not paying the rent deposit back after the end of the leasing relationship. So far, the case seemed to be a typical conflict between landlord and tenant. What made the case interesting for newspapers and other media is the reason the landlord brought up for not paying the rent deposit back. The landlord could prove that the marble floor in the bathroom had been damaged. He furthermore could prove that the floor was damaged because the male tenant was standing up when he was urinating (instead of sitting down which has become more common in German and other countries' households). The Local Court Düsseldorf now had to decide if the use rights of (male) tenants include the right to urinate standing up. The Court held that, although sitting down for urinating has become more common in German households, until that point the custom of urinating standing up was still widespread. A man who stands while urinating (so held the Court) would probably encounter conflicts with female roommates, but could not predict the damage to the marble floor (Düsseldorfer Amtsgericht 2015: para. 38). The landlord had to pay the rent deposit back. The German urinating case is remarkable not just because a court had to position itself on an issue that is usually negotiated between men and women on other discourse levels. The case is also remarkable because it treats a certain need as self-evident to be claimed through housing. The right to housing, one could conclude, includes the unspoken right to use a toilet in privacy—including the choice of how to use the toilet. Another prominent subject that German courts had to decide against the backdrop of tenants' rights was smoking (Landgericht Düsseldorf 2016). In general, German tenants can smoke in their flat, if they do not disturb the peace in the house and violate the bid of consideration (German: Gebot der Rücksichtnahme) (Landgericht Düsseldorf 2016: para. 40).

This chapter has shown that there are several actors monitoring the implementation of the human right to housing. Moreover, it has implied that the global discourse on housing jurisprudence is dominated by Western scholars who look upon the national jurisprudence either in Western countries or the European jurisprudence (including the complaints submitted under the RESC). Regarding housing through the lens of the law means to consider and balance different rights and duties. Depending on the circumstances, the right to housing could comprise of:

- the right to formal property,
- the right to informal property,
- the right to a home,
- the right to privacy,
- the right to family life,
- the right to identity,
- the 'right' to urinate standing up,
- the 'right' to smoke, and
- many more.

Depending on the circumstances, the right to housing could also comprise:

- the states' duty to respect, protect, and fulfill this right and all other rights that are interrelated with housing;

- the landlords', developers', local authorities', and welfare organizations' duty to support the state;
- the duty of the CESCR, the ECSR, FEANTSA, the Special Rapporteur on inadequate housing, and national and international courts to monitor the implementation of this right and all other rights that are interrelated with housing;
- the tenant's duty not to disturb the peace in the house they live in and violate the bid of consideration; and
- the opportunity of NGOs, other organizations, and individuals to claim a violation of this right and all other rights that are interrelated with housing.

Usually, there are several parties involved when it comes to balancing the rights, duties, and opportunities under the umbrella of the human right to housing. Certainly, the opportunities, as well as relationships between the rights and duties differ from case to case and from country to country. Moreover, this chapter can only indicate what the next chapters, particularly Chapters 5 and 6, will prove: The rights, duties, and opportunities under the umbrella of the human right to housing have always been and still are changing.

3.2 Housing and Discourse Research

When scholars examined housing, they observed several aspects. Some scholars discussed the advantages and disadvantages of homeownership in relation to renting and vice versa (e.g., Czasny et al. 2008). Other scholars discussed if and why housing is an issue of welfare policy (e.g., Matznetter & Mundt 2012) or analyzed the connection between homeownership and welfare policies (e.g., Stamsø 2010). Some scholars observed certain groups affected by inadequate housing such as slum dwellers (e.g., Davis 2006), informal settlement residents (e.g., B. Davy & Pellissery 2013), or homeless people (e.g., Steinmeier 1992). Others observed the role of the state, the market, and other actors (Edgar et al. 2002). Scholars examined housing on a local level and discussed neighborhoods (e.g., Galster 2012) or the relationship between tenants and landlords (e.g., Bierre et al. 2010). Some compared the housing policies of different countries with each other (e.g., Hoekstra 2010). Others concentrated on housing policies of complex global players such as the World Bank (e.g., Pugh 1991 and 1995) or UN-Habitat (e.g., Gilbert 2007).

This section discusses housing as a topic of discourse reserach. However, other planning issues have also been analyzed through discourse analysis (e.g., Glasze & Mattissek 2009; Günzel 2016; Kolocek 2009; Mattissek 2008). Jago Dodson (2007) examined the production of housing policy knowledge by policymakers and scholars. The author observed the housing policies of the United States, the United Kingdom, the Netherlands, and particularly of New Zealand and Australia. Dodson (2007: 52) regarded these countries' jurisdictions, policy statements, reports, and documents concerning the housing apparatus. He aimed to examine 'the extent to which the objects and subjects of housing policy are the conceptual artifacts of the discourses that are articulated by housing officials' (Dodson 2007: 2). Dodson's (2007: 25–57) methodological approach was based on the philosophical school of radical empiricism; he was particularly interested in the state's role in the process of knowledge production. The author examined the relationship between policy and truth (Dodson 2007: 26–27). He assumed

> that the methodological direction established by Foucault and others provides a means of developing methods that can assist to identify how concepts and ideas about housing and housing relationship, in specific historical periods, come to constitute housing policy, and how these notions are translated into practice. (Dodson 2007: 39)

Dodson stated that housing scholars and policymakers do not merely solve problems related to housing, but are in central positions to articulate what housing problems are: 'Government from this perspective thus acts to produce the effects of its own discourse' (Dodson 2007: 46). To make his point, the author brought up homelessness:

> In housing policy, the condition of homelessness is not present in a discursive sense, until the institutions of housing policy initiate a

> pragmatic procedure to articulate this condition, constitute it as visible, and make it able to be acted upon, via various policy measures and practices. (Dodson 2007: 46)

Dodson (2007: 255) identified three substantive discourse transformations. The first one took place in the mid-to-late nineteenth century when reformers claimed that the inadequate housing conditions and the spread of slums emerged through the high expansion of industrial cities. Dodson identified the early to mid-twentieth century as the second transformation, which strengthened the state's self-perception as the provider and allocator of housing for the poor. The contemporary period is marked by the last transformation: 'The capacity of the state to perceive and comprehend housing problems has been placed in doubt, often by the state itself' (Dodson 2007: 255). Dodson (2007: 258) could prove that non-state actors like landlords, housing associations, or community housing organizations have often taken the role of housing providers (Dodson 2007: 255), albeit not in the full scope: '[N]o government policy agent has seriously proposed the abrogation of all government interest of concern for housing' (Dodson 2007: 256). Despite the more theoretical background on the methodological level, Dodson's study has important similarities with the present investigation. Dodson's focus was on the government's perspective. He articulated the power of the government to define what problems are relevant to housing and what problems are not.

Other scholars have also regarded housing issues from discourse analytical points of views. Bierre et al. (2010) observed the identities of landlords and tenants in the private rental market of New Zealand. Their study used methods of Critical Discourse Analysis (CDA) and was guided by Habermas's theory of communicative action. By analyzing key-informant interviews and housing policy documents, they identified two stereotypical identities, the 'Ma and Pa landlord' and the 'risky tenant' (Bierre et al. 2010: 31–33). Gilbert, by analyzing the use of the term slum in UN publications, particularly in the Cities without Slums Initiative (e.g., UN-Habitat 2003), asserted:

> The danger with the term 'cities without slums' is that it is just a slogan; rhetoric that carries with it an empty promise. What worries me too is that

> use of the word slum will recreate many of the old stereotypes about poor people that years of careful research has [sic] discredited. By using an emotive word, the UN draws attention to a real problem but, in doing so, it evokes a response that it cannot control. (Gilbert 2007: 710)

Other housing issues that have been analyzed through discourse analysis are, to name a few, refugees and asylum seekers (Every & Augoustinos 2007; Lynn & Lea 2003) or the politics of race and space in South Africa (Dixon et al. 1994). Homelessness was a subject of several research approaches (Fopp 2009; Forte 2002; Huckin 2002; Meert et al. 2004; Penner & Penner 1989; Torck 2001).

Fopp (2009) discussed the three metaphors safety net, career, and pathways in homelessness research and Australian policy documents. Although the author acknowledged that a metaphor can have a diversity of meanings, he concluded that the three metaphors 'are not necessarily neutral ways of examining homelessness and the experience of people who are homeless' (Fopp 2009: 289). In the case of the word career, for instance, the author claimed that using such a word with respect to homelessness neglects the structural causes of homelessness:

> Although there may be 'chequered' careers, or a step in the process may be a 'bad career move', it is worth emphasizing that, socially understood, careers are about choices, developing potential, positive evaluation, opening opportunities and securing rewards. Homelessness, on the contrary, is socially perceived as a downward and retrograde step. (Fopp 2009: 283)

In a cross-cultural exploration, Torck (2001) examined different street newspapers from four countries: the Netherlands (Z-magazine), the United Kingdom (The Big Issue), France (L'Iternérant), and the United States (Street Sheet). He concluded that, in contrast to the Street Sheet in the United States, the three European street newspapers gave a limited political platform to homeless peoples' voices and focused more on personal narratives and poetry. Penner and Penner (1989) observed societal perceptions of homelessness by analyzing cartoon media. They found out that cartoonists frequently described homeless people from a stereotypical perspective: a middle-aged man in old baggy clothes, sitting with a bottle by an open fire (Penner & Penner 1989: 101). The heterogeneity of homeless people, as the authors concluded, is too complex for cartoon depiction (Penner & Penner 1989: 105). Meert et al. (2004) examined homelessness in the written press. They found out that homelessness is a highly seasonal issue that often comes up in the cold weather months; the newspapers represented dramatic and polemic aspects of homelessness and often ignored the discussion of structural causes (Meert et al. 2004: 34).

All in all, housing has been a subject of several discourse analytical studies, albeit with different focal points. This chapter has demonstrated at least two main things: First, as housing is often regarded as an issue of the states' responsibility, the states are in the central position to articulate what housing problems are. Second, examples of inadequate housing such as slums or homelessness are emotive terms, often in danger of being regarded from a stereotypical point of view. However, until today, no study has examined housing through discourse analysis in a global comparative context.

3.3 Homeownership and Renting

The relationship between social policy research and housing research is somewhat strange. In 1987, Torgersen described housing as the 'wobbly pillar under the welfare state'. The author discussed the institutional peculiarity of housing as part of the welfare state (Torgersen 1987: 116). Housing as a commodity, he stated, differs from other goods (Torgersen 1987: 118). The metaphor of the wobbly pillar has been picked up in housing discourses (e.g., Benjaminsen & Dyp 2008: 47; Edgar et al. 2002: 8; Kemeny 2001: 54; Stamsø 2010: 64; Stephens & Fitzpatrick 2007: 206). Kemeny (2001: 54) noticed that housing differs from the other pillars of the welfare state (social security, education, and healthcare) for a few reasons: It is characterized by its high capital intensity and rarely considered as a universal form of public provision. The author claimed that housing is still strikingly absent from comparative welfare research (Kemeny 2001: 56). He regarded housing as the key to understanding why welfare states differ from one another so much: '[C]hange housing in important respects and the consequences of that change are likely to reverberate through the whole welfare system' (Kemeny 2001: 57). As Clapham noticed, housing should be regarded as an instrument of social policy 'in its own rights but also as a means of achieving objectives in areas such as health and poverty'(Clapham 2012: 164). I think, what Kemeny frustrated is not that housing scholars did not take interest in welfare policy—because they did—but that prominent welfare theories seldom put housing as a pillar of the welfare state on the agenda.

Housing is one of the most prominent of the ESC rights (Craven 2002: 329) and at the same time a complex commodity (Oxley 2004: 8). It is 'variously connected to infrastructural services, land policy and land management, to capital market and financial systems, and to macroeconomic conditions' (Pugh 1995: 80). Scholars have developed several indicators to measure the quality of housing, in particular in Western countries (e.g., EUROFOUND 2009). In the following sections, I will differentiate between the two main tenure forms renting and homeownership. When looking at the homeownership rate of a country, the first impulse is to conclude a high welfare status if the rate is high, meaning that more homeowners stand for more wealthy people. However, some researchers stressed that it is quite the opposite, for example, in developing countries:

> In many developing countries, home ownership rates are often high because of weak legal systems, underdeveloped land-use and the rural-family base of economic production. Industrialization and urbanization tend to stimulate the growth of rental housing, necessary to support growing populations of workers, and the development of recognizable 'systems' of housing which involve a complex coordination of diverse areas of planning, legislation, financing, construction, marketing, distribution and consumption, among other things. (Ronald 2008: 5)

In the early 1980s, Kemeny (1980) stated that in societies with low public retirement pensions and poor public welfare provisions for the elderly, people were forced into homeownership as a private provision for their old age. Kemeny regarded the rate of homeownership as an indicator of poor welfare provisions. The author (Kemeny 1980: 377) stressed that countries with a well-organized and strong labor movement have developed larger and more attractive cost rental sectors as an alternative to owner occupation. Kemeny (1981: 145) heavily criticized the idea that homeownership in English-speaking countries was widely regarded as a good thing—which it was not. Based on case studies in Australia, Sweden, and Great Britain, he claimed a systematic discrimination in favor of homeownership that, to Kemeny, had far-reaching consequences:

> It restricts real choice in housing tenure; it ossifies tenure patterns by discouraging two-way movement into and out of different forms of tenure, it stratifies housing tenure in terms of social class; and, by artificially stimulating the expansion of home-ownership, it amplifies the limitations of home-ownership as a form of tenure. In addition, policies to encourage home-ownership are necessarily inequitable, and [. . .] the attempt to force as many households as possible into home-ownership creates acute problems of its own. (Kemeny 1981: 145)

Kemeny (1981: 146) recommended a tenure-neutral housing policy in which the governments could encourage the development of a wide range of types of tenures. Public renting, to Kemeny, should converge into a cost-rental sector by allowing public housing to compete with private landlords for tenants. For the consumer, 'the distinction between "private" and "public" renting would therefore disappear' (Kemeny 1981: 146). In a self-critical review of his earlier work, Kemeny (2005) referred to a statistical analysis of Castles (1998), who regarded homeownership as an investment over the life cycle and asked whether homeownership might make a difference to the understanding of welfare outcomes. Castles regarded retired people as one category of welfare recipients for whom homeownership is crucial because they have escaped from the labor market. The assumption was as simple as convincing: 'When individuals own their own homes, they can get by on smaller pensions' (Castles 1998: 12–13). Castles and Ferreira (1996: 170) found out that many OECD countries with a high homeownership rate—Finland, Norway, Spain, and the

English-speaking countries Australia, Canada, Ireland, New Zealand, and the United States—have low protection expenditures for the aged. They called it 'the really big trade-off' between owner occupation and pensions. OECD countries with a low homeownership rate—Austria, Belgium, Denmark, Germany, Luxembourg, the Netherlands, and Sweden—often have high expenditures for the age (Castles and Ferreira 1996). The real estate crises in the United States and Europe have, however, brought homeownership as a cornerstone of private welfare provision into question. Regarding homeownership as an investment over the life cycle (Castles 1998: 6) is quite puzzling. Private homeownership understood as the 'commodification of housing' (Kemeny 1980: 384–385) seems to de-commodify labor because it reduces the necessity and the level of retirement pensions—which Esping-Andersen (2011, 4.3) regarded as one of three indicators for measuring the de-commodification of labor. Castles (1998: 17) and Kemeny (2005: 74), however, agreed that more research on the relationship between owning and renting and welfare provision is necessary.

The relationship between tenure forms and housing satisfaction is even more confusing. Karl Czasny and his research team from the SRZ (Stadt + Regionalwissenschaftliches Zentrum) observed the level of housing satisfaction—they called it satisfaction with home—in Austria and other EU member states (Czasny et al. 2008). They found out that house and flat owners in Austria are more satisfied with their housing situation than Austrian tenants are (Czasny et al. 2008: 26). The researchers analyzed data of the Eurobarometer, a European public opinion survey. They compared European countries with a high homeownership rate (like Spain, Ireland, and Hungary) with countries with a low homeownership rate (like Austria, Germany, and the Netherlands). Surprisingly, the rate of people who are satisfied with their housing situation was, on average, higher in countries with a small ownership rate than in countries with a high ownership rate (Czasny et al. 2008: 19).

Based on the European Community Household Panel (ECHP), Elsinga and Hoekstra (2005) examined the relationship between the type of tenure and housing satisfaction in the eight selected EU countries of Ireland, United Kingdom, the Netherlands, Denmark, Austria,

Greece, Italy, and Spain. They also differentiated between homeowners and tenants. Their findings are similar: In all eight countries, homeowners were more satisfied with their housing situation than tenants were (Elsinga & Hoekstra 2005: 414). The scholars did not discuss the differences between countries with a high ownership rate and countries with a lower one in detail. In the countries of Southern Europe (Italy, Spain, Greece), the housing satisfaction is lower than in the other five countries (Elsinga & Hoekstra 2005: 414). The findings are confusing. How could one explain that, in a European comparison, people from countries with a low homeownership rate are more satisfied with their housing situation—although in every country homeowners are more satisfied than tenants?

From a mathematical viewpoint, the findings are less paradoxical than it appears. I would like to clarify the paradox with a simple hypothetical model. The model assumes two countries; each country has 1000 inhabitants, which will be titled as Owners' Country and Tenants' Country for this purpose. In Owners' Country, 700 people are homeowners and 300 are tenants. The relationship in Tenants' Country is vice versa: 700 tenants and 300 homeowners. In both countries, the homeowners are more satisfied with their housing situation than the tenants. On a ten-point scale (1 point = absolutely unsatisfied, 10 points = fully satisfied), the homeowners value their housing satisfaction in the Owners' Country with 3 points, the tenants value it with 2 points (Table 3.1). In Tenants' Country, the 300 owners are almost fully satisfied with their housing situation (9 points), and the tenants are less satisfied (1 point) than the tenants in the Owners' country.

Table 3.1 demonstrates that, either way, the population in Tenants' Country is, as a whole, more satisfied with the housing situation (3.4 points on average)[1] than the population in Owners' Country (2.7 points on average).[2] The key explanation lies in the high satisfaction rate of the 300 homeowners in Tenants' Country.

[1] (300 * 9 + 700 * 1) / 1000 = 3.4.

[2] (700 * 3 + 300 * 2) / 1000 = 2.7.

Table 3.1 The ownership/satisfaction paradox

	Inhabitants		Housing satisfaction		
	Homeowners	Tenants	Homeowners	Tenants	On average
Owners' country	700	300	3 points	2 points	2.7 points
Tenants' country	300	700	9 points	1 point	3.4 points

The findings of Czasny et al. (2008) and Elsinga and Hoekstra (2005) should be considered carefully, as the information represents correlation rather than causality. Elsinga and Hoekstra admitted for good reasons that such results cannot clarify the causality: 'If homeowners appear to be more satisfied than tenants, one can only guess if this is due to the characteristics of the tenure, the effects of policy, cultural influences or some other factors' (Elsinga & Hoekstra 2005: 410). Additionally, many more tenure forms exist besides ownership and renting (Torgersen 1987: 123–125; UN-Habitat & GLTN 2008). Elsinga and Hoekstra (2005: 405) also noticed that theories that emphasize the benefits of homeownership are usually formulated by scholars from English-speaking countries. Taking Kemeny's (1980) early assumption seriously that the welfare provisions in countries with a low homeownership rate (a Tenant's Country) are better, one could guess why homeowners are so lucky when they live in a Tenants' country. They profit from both the good welfare provisions and their homeownership.

The discussion has indicated that, on the one hand, countries differ by the tenure form they prefer, and on the other hand, that scholarly discourse shows no consensus concerning the best tenure form. What the different positions and tenure traditions have in common is that they all regard the existence of a market, whether it is a renting or a homeownership market, as self-evident. This, however, does not apply to millions of people who live outside of formal housing markets, as the next section will show.

3.4 Homelessness and Spaces of Inadequate Housing

The many forms of inadequate housing differ. For the present book, they are separated into two forms: Homelessness and Spaces of Inadequate Housing, which cover slums, informal settlements, illegal settlements, and many other examples.[3]

Homelessness is a special form of inadequate housing because homeless people neither own nor rent—they occupy. From the perspective of law, a homeless person, for instance, has many rights on different levels, including the mentioned ICESCR, several other international covenants and conventions (Leckie 2001: 151–153; UN-Habitat 2009a: 11), and the national law. Homelessness entails a violation not only of the human right to housing, but it can be regarded as a violation of many human rights (Lynch & Cole 2003). The existence of housing rights does not automatically mean that states have the obligation to immediately eliminate homelessness (Craven 2002: 330). But human rights seem far away when we consider a homeless woman collecting deposit bottles or begging in front of a railway station. Homelessness is often understood to mean having no belongings, and therefore, the issue is seen as a property problem (Baron 2004). Waldron (1991) discussed why homeless people are unfree, namely because they have no property. Homelessness could also be a rural problem (Milborne & Cloke 2006), but most homeless people sleep in urban areas that are often near or at places of high economic value. The land beneath a homeless person's feet is often much more valuable than the land beneath a single-family house at the edge of a city. Thus, urban homelessness often stands for the spatial proximity of extreme poverty and wealth.

[3] I started thinking about the overlaps and differences between certain forms of inadequate housing some years ago. Section 3.4 is drawing from a chapter that I wrote in an early version in Kolocek (2012). I tried out the forms (codes) of homelessness and Spaces of Inadequate Housing in different contexts. I examined both codes in the States Parties reports from across Latin American and African ICESCR member states (Kolocek 2012) and in the States Parties reports of ICESCR member states from the European Union (Kolocek 2013). Moreover, I discussed the development of both codes from a methodological point of view in Kolocek (2014).

Despite the stereotypical perception of homelessness that many people have (Meert et al. 2004; Penner & Penner 1989), alternate definitions of homelessness exist in scholarly literature (e.g., Baron 2004; Springer 2000; Waldron 1991). Frequently, researchers have criticized the countless definition attempts that often bear misunderstandings (O'Flaherty 1998: 9). These misunderstandings have consequences for investigating and combating homelessness, particularly on a global level (Springer 2000: 477), or, in Neale's words: 'If policy cannot define homelessness, how can it hope to respond to it?' (Neale 1997: 55). As O'Flaherty (1998: 18–19) noticed, the numbers of homeless people can vary vastly depending on the definition one's perspective is based upon. Society often seems to have a stereotypical perception while many researchers agree that homelessness goes beyond this stereotypical perception:

> In the minds of many people, homelessness is identified with the condition of tramps and vagrants. Research results are often received with skepticism. 'Where are those hundreds of thousands of homeless people? If they are not in the street, they are not homeless', is a typical reaction. (Avramov 1995: 71)

Daly (1996: 247) stressed the significance of the relationship between language und action when studying homelessness. He compared the homelessness policies of Great Britain, Canada, and the United States and emphasized the connection between a nation's definition and its politics. Daly (1996: 25) distinguished between a narrow and a broad definition: The United States has a narrow definition, Great Britain has a broad definition, and Canada's take lies somewhere between these two definitions. At present, a consensus has been reached that homeless people are a heterogeneous group.

In 2005, FEANTSA presented a European typology of homelessness and housing exclusion (Table 3.2; for a critique of its validity, see Amore et al. 2011). In this typology, people were classified as belonging to one or more of four conceptual categories: roofless, houseless, insecure, and inadequate; and 13 different operational categories (hereafter: oc).

Table 3.2 European typology of homelessness and housing exclusion

Roofless	People living rough (operational category 1)
	People living in emergency accommodations (operational category 2)
Houseless	People in emergency accommodations for the homeless (operational category 3)
	People (women) living in women's shelters (operational category 4)
	People in accommodation for immigrants (operational category 5)
	People due to be released from institutions (operational category 6)
	People receiving longer-term support (operational category 7)
Insecure	People living in insecure accommodations (operational category 8)
	People living under the threat of eviction (operational category 9)
	People living under the threat of violence (operational category 10)
Inadequate	People living in temporary non-conventional structures (operational category 11)
	People living in unfit housing (operational category 12)
	People living in extreme overcrowding (operational category 13)

Source: FEANTSA (2005). Abridged version.

The operational categories illustrate that the boundaries between homelessness and other forms of inadequate housing are not clear. This is not only a challenge for scholarly literature, but a phenomenon of everyday life. A homeless woman may sleep in a park in the summer time and visit an overcrowded (oc 13) emergency accommodation (oc 3) in the winter, where she could be under the threat of violence (oc 10). The relationship between homelessness and other forms of inadequate housing is even more complex. The eviction from a slum can cause homelessness. An inadequate night shelter can be an answer to combat homelessness that nevertheless remains a place of inadequate housing.

Frequently, the causes of homelessness are separated into two groups: individual and structural causes. Examples of individual causes are mental illness, alcoholism, substance abuse (Main 1998: 42), or a traumatic event in the individual's life (Avramov 1995: 80). Structural causes emphasize aspects like the labor (or housing) market situation, natural disasters, social politics, or poverty. Main (1998) criticized that researchers often focused on one of these groups and neglected the other one. Neale (1997) made a similar argument. She tried out other more theoretical explanations (feminism, post-structuralism, postmodernism, structuration, and critical theory) to understand homelessness and its causes (Neale 1997). Tipple and Speak (2009) distinguished homeless people living in industrialized countries from homeless people living in developing countries. Regarding the structural causes, they made the interesting assumption that '[in] many industrialized countries, homelessness has little directly to do with housing shortages. [...] In developing countries, however, housing supply shortfalls in absolute terms are undoubtedly a structural cause of homelessness' (Tipple & Speak 2009: 33).

The complexity of causes and the heterogeneity of the group of homeless people lead to many different answers about how to respond to homelessness. Countless policies, programs, campaigns, and strategies on different levels exist to combat or prevent homelessness, such as the International Year of Shelter for the Homeless 1987 campaign (UN 1981) or a single soup kitchen in the city of Hamburg. On the FEANTSA website (www.feantsa.org), for instance, there are links to different national strategy papers of European countries as well as papers from Australia and the United States. Every one of these policies, programs, campaigns, and strategies is again separated into different measures on different levels. Policies on the housing market (O'Flaherty 1998) can cause or, conversely, prevent homelessness, but are only indirectly connected to homelessness. Due to the interdependence of homelessness, even developments in other sectors—for example, the labor market or the health sector—can influence homelessness. Actors can respond to homelessness in countless different ways, for example, through the building of new housing units, financial measures, health care, rehabilitation programs, social support, and more. The measures differ in several veins. Some are more preventative; others focus more on reintegration goals. Some measures are large-scale programs (like a new national building program) that respond to housing problems of thousands of people; others could be regarded as small-scale measures like the above-mentioned soup kitchen.

Street children live in a special form of homelessness, albeit not every street child is homeless:

> Some street children are part of entire families who live on the street. Others are born to older streetgirls. Some street children are 'on the street', which means that they still see their families regularly and may even return every night to sleep in their family homes. Children 'of the streets,' on the other hand, have no home but the streets. Even if they occasionally spend time in institutions for children or youths, they consider the streets to be their home. (WHO 2000: iii)

For the present research, however, street children are treated as a subcode of homelessness. I will examine how much the states' way of speaking changed when they explicitly addressed street children.

Obviously, being homeless means living in a situation of inadequate housing, even though, homelessness is only one example of inadequate housing. Other examples are slums, informal settlements, illegal settlements, shanty settlements, irregular settlements, pirate settlements, unauthorized communities, emergency shelters, refugee camps, and many more. Some countries use their own terms, like barrio (Spanish), bidonville (French), favela (Portuguese), gececondu (Turkish), Elendsviertel (German), trushchobi (Russian) or hood and ghetto (American English) (UN-Habitat 2003: 9–10). For simplification, my term to summarize these inadequate housing forms is: **Spaces of Inadequate Housing** (Kolocek 2012 and 2013). Using the umbrella term Spaces of Inadequate Housing should not be understood as neglecting the heterogeneity of each of the forms that are summarized. As Gilbert (2007: 702) stated, even the term slum is not homogenous and can denote a range of situations.

People living in Spaces of Inadequate Housing sometimes own and sometimes rent; they sometimes participate in formal markets and sometimes informal. There is a key difference between both forms. Slums, informal settlements, or emergency shelters are spaces where people affected by inadequate housing live. The spaces are the primary focus in this context, and the people secondary. Homelessness, however, stands for people who are affected by inadequate housing. Homeless people are the primary consideration within that context and only then does the question come up of which spaces they occupy. Many definition approaches exist for slums and squatter settlements as well:

> Slums and squatter settlements are difficult to separate. However, slums generally refer to housing, regardless of tenure, which has fallen into such disrepair that it constitutes a general condition for a neighbourhood or community. A squatter area could also be a slum. (Aldrich & Sandhu 1995: 19)

Slums (or informal settlements), as a relative concept with many local variations and different sizes, are too complex to be encompassed in one global definition (UN-Habitat 2003:11). UN-Habitat worked out a simple and broad definition of a slum. A slum is 'a settlement in an urban area in which more than half of the inhabitants live in inadequate housing or lack basic services' (UN-Habitat 2006: 21). UN-Habitat suggested focusing on the household as the basic unit of analysis: 'A slum household is a group of individuals living under the same roof in an urban area who lack one or more of the following conditions' (UN-Habitat 2006: 21). These conditions are durable housing, sufficient living area, access to improved water, access to sanitation, and secure tenure. In addition, UN-Habitat (2003: 80–85) distinguished between slums of hope, which are progressing settlements, often in a process of improvement, and slums of despair, which describes declining neighborhoods. Other categories used to differentiate slums are origins and age, location and boundaries, size and scale, legality and vulnerability, and different developmental stages (UN-Habitat 2003: 85–95).

Many scholars had third world countries in mind when they considered Spaces of Inadequate Housing. This perception seems to be changing, albeit slowly. In 2009, the United Nations Economic Commission for Europe presented its first publication concerning informal settlements for the Europe Region (ECE 2009). This was the first time that the challenge of informal settlements gained a spot on the European agenda. The study pointed out that in over 20 countries of the UNECE region, more than 50 million people live in informal settlements (ECE 2009: xv). Moreover, the study reviewed different policy interventions that are also known from developing countries, such as legalization, regularization, and upgrading, or resettlement and relocation (ECE 2009).

Problems with measuring Spaces of Inadequate Housing also stem from contradicting definitions and emotional connotations. Statistics concerning Spaces of Inadequate Housing are often more than simple undemonstrative pieces of information about emotionless numbers. Inadequate housing is closely connected with questions of rights—legal rights as well as everybody's own understanding of social justice. The media, as well as academics, have often used statistical information to emphasize the significance of the problem like Kofi Annan did in the foreword of the report 'The Challenge of Slums' (UN-Habitat 2003). I will not discuss the vast amount of statistical information in detail. Nevertheless, there is no doubt in scholarly discourse that more than one billion people throughout the world fall under my main two forms of inadequate housing (UNDP 2014), and the number is, in fact, increasing (Kothari & Chaudhry 2012). What are the causes for Spaces of Inadequate Housing? In one sentence:

> [S]lums [and other Spaces of Inadequate Housing] develop because of a combination of rapid rural-to-urban migration, increasing urban poverty and inequality, marginalization of poor neighborhoods, inability of the urban poor to access affordable land for housing, insufficient investment in new low-income housing and poor maintenance of the existing housing stock. (UN-Habitat 2003: xxxviii)

Slums, for example, spread because current slum dwellers build millions of huts and shacks. Nevertheless, by regarding a slum as a problem, the focus lies on structural (planning) failures. The most mentioned cause is urbanization, in other words, rural-urban migration (Davis 2006: 1–19). This does not mean that Spaces of Inadequate Housing are necessarily geographically close to inner cities, but they are on the periphery of Third World Cities (Davis 2006: 37). However, urbanization has its own causes and explanation models. Aldrich and Sandhu (1995) distinguished between four models of the process of urbanization: A basic demographic model emphasizes the push and pull factors of the migration to cities; a hegemony model evaluates trade relations between industrialized and Third World countries; a dependency model focuses on the national economy and the dynamics of an international market; and an econometric model looks at internal and external factors like the gas and oil prices and regards the informal sector as part of the economy (Aldrich & Sandhu 1995: 23–26). Another reason for the spread of Spaces of Inadequate Housing is the failure of governance based on the lack of political will (UN-Habitat 2003: 5–6). Poverty is regarded as both a cause and an effect (UN-Habitat 2003: 28–31).

Global, national, or local actors try to improve the housing situation of people living in Spaces of Inadequate Housing with different constructional measures. For example, they improve infrastructural aspects (including access to water, sanitation, and energy), or they build new or better houses. These measures are often rooted in upgrading, slum improvements, or urbanization programs. Other measures are resettlement or relocation. Resettlement and relocation measures were often regarded critically due to the destruction of social networks (UN-Habitat 2003: 150). Another important measure is the protection from eviction. Just like homeless people, dwellers in Spaces of Inadequate Housing sometimes profit from different social support measures, such as nutrition, health care, or education programs. Housing credits, loans, microcredits, or subsidies are typical financial measures. Land titling is probably one of the most discussed responses to Spaces of Inadequate Housing.

To sum up, both main forms—homelessness and Spaces of Inadequate Housing—have some common ground. People living in Spaces of Inadequate Housing and homeless people live in all regions of the world. Homelessness is not only a problem of Western countries nor do Spaces of Inadequate Housing dwellers only live in countries of the global South. Millions of Spaces of Inadequate Housing dwellers and homeless people all around the world often live outside of formal housing markets. At present, both forms are more often regarded as an urban rather than a rural phenomenon.

The key difference becomes visible when taking the spatial dimension —the land—into account. Homeless people, despite having economically valuable land beneath their feet, are usually regarded as poor persons who need help. Scholarly as well as political discourses only seldom regard homeless people from an economic point of view. In the perception of many, a homeless person is often a helpless victim that must be carefully integrated back into society. In such an argument, the term society is often implicated (albeit unspoken) as nothing other than formal housing and labor markets. The main difference is that while homeless people often occupy places of potentially high economic value temporarily, people living in Spaces of Inadequate Housing usually occupy spaces for a long time, causing the land beneath their feet to somehow belong to them, illegally or informally. The land titling approach indicates that the 'dead capital,' understood as the land beneath the people's feet and the people's manpower, should be woken up to integrate them into society (Section 4.2).

This differentiation shows that the thinking about homelessness is closely related to thinking about roofless and houseless people. Thinking about other forms of inadequate housing, for example, imagining a slum, is more likely connected with the location where the slum dweller lives. A simple test helps to recognize this key difference. The reader might try to see in his or her mind's eye a picture of the phenomenon homelessness—without a homeless person. This hardly works. On the other hand, imagining a picture of a slum—without a slum dweller— would probably work (better).

This chapter has introduced the broad field of housing research and shown that discourses on housing differ, depending on different aspects. In the context of Western welfare policies, scholars either ignore housing issues, or they talk about the tenure question and ask how to influence formal housing and land markets. Homelessness and Spaces of Inadequate Housing such as slums or informal settlements prove that many people live outside of such markets, informally or illegally.

4

Land Policy Meets Social Citizenship on a Global Level

This chapter discusses why and how social policy became more global during the last few decades (Section 4.1). It starts by introducing global social citizenship, which stands both for a research discipline that analyzes the global character of social policy and for a leitbild that puts the individual including his or her social rights in the center of attention. Section 4.2 points out how land policy works as social policy and presents the discourse on land titling and informality (including a short excursion about the World Bank's housing policies). While proponents of land titling regard land as a commodity, proponents of informality claim that other arrangements—based on trust, security, stability, protection, and control—are at least sometimes more suitable than the integration of people and land in formal property markets. In the last section (Section 4.3), I first present concepts to measure social policy. Second, I show how housing scholars picked up Esping-Andersen's (2011) concept of de-commodification of labor and developed different approaches of de-commodification of housing. Based on the theoretical discussion of the role of land in Section 4.2, I then generate a theory of de-commodification of land use that responds to

M. Kolocek, *The Human Right to Housing in the Face of Land Policy and Social Citizenship*, DOI 10.1007/978-3-319-53489-3_4

the main argument of the book, namely that policymakers around the world will not be able to completely fulfill the human right to housing.

4.1 Global Social Citizenship

In simple terms, global social citizenship is about welfare and social policy in a global context. I regard welfare policy and social policy as the same thing in so far as I do not differentiate whether an author calls his or her theory a welfare policy theory or a social policy theory (or a welfare model, position, etc.). T.H. Marshall defined welfare as 'a compound of material means and immaterial ends [that] is located somewhere on the axis which runs between the poles of wealth and happiness' (Marshall 1965: 261). Welfare theories or social policy theories try to explain how the states—or other actors—try to make the people wealthier and happier, or at least not poor and unhappy.

Social policy is 'both a scholarly activity and the actual practice of governments and other agencies that affect the social welfare of populations' (Deacon 2007: 1). The scholarly activity has its roots in the Western political philosophy of rich industrial and post-industrial countries (Gough & Wood 2004: 3). Deacon named three approaches to characterize social policy: social policy as sector policy; social policy as redistribution, regulation, and rights; and social policy as policy issues (Deacon 2007: 4–5). The third approach understands social policy along the lines of what social policy scholars do. What social policy scholars usually do at the beginning of their books and papers is that they quote T.H. Marshall's (1950) Essay 'Citizenship and Social Class.'

> [F]rom the moment that Marshall first delivered his famous lectures in 1949, *Citizenship and Social Class* and citizenship as a social science became virtually one and the same. It is hard in today's context where a thousand concepts bloom to remember that in the absence of any serious preceding sociological heritage, Marshall essentially invented the social approach to citizenship. Once he did, he became until recently its 'sole owner'. (Somers 2008: 151–152, emphasis in the original)

T.H. Marshall explained how citizenship that emphasizes equality could grow and flourish side by side with social class, which is a system of inequality (Marshall 1950: 29). The equality of citizenship means the equality of status (rights) while rising class differences emerge from a variety of factors related to the institutions of property and education, and the structure of the national economy (Marshall 1950: 31). Citizenship can be understood as a status that makes a person a full member of a community—or, in other words, citizenship is the right to have rights on both the empirical and the normative level (Somers 2008). In Marshall's words: 'All who possess the status are equal with respect to the rights and duties with which the status is endowed' (Marshall 1950: 28–29). Citizenship has three basic elements: The civil element comprises liberty of person; freedom of speech, thought and faith; the right to own property and to conclude valid contracts; the right to justice; and even economic rights like the right to work. The political element means above all the right to participate in the exercise of political power. By invoking the social element, Marshall meant 'a whole range from the right to a modicum of economic welfare and security to the right to share to the full in the social heritage and to live the life of a civilised being according to the standards prevailing in the society' (Marshall 1950: 11). Marshall described citizenship on the empirical level and observed its development during the last centuries. He stated that each of the three basic elements (bundle of rights) developed in a certain century: Civil rights in the eighteenth, political rights in the nineteenth, and social rights in the twentieth century (Marshall 1950: 14), even though there was a considerable overlap between political and social rights (Marshall 1950: 21). As the present research concentrates on housing as an issue of social policy, the civil and political rights will not be considered in more detail. Instead, I will focus on social citizenship in the context of housing. In a contemporary reading, social citizenship

> means that the wealthy and the poor are treated as full members of society [. . .]. Social citizenship also means, however, that the wealthy have the right to be unequal—to possess more and better functionalities than

> others—as long as the poor receive their right to a civic minimum in exchange for equality. (B. Davy 2012: 211)

The original source of social rights, to Marshall (1950: 21), was membership of local communities and functional associations that were progressively replaced by a system of wage regulation and the Poor Law (in England). Marshall regarded the Poor Law (in England) as an example of the divorce of social rights from the status of citizenship: 'The Poor Law treated the claims of the poor not as an integral part of the rights of the citizen, but as an alternative to them—as claims which could be met only if the claimants ceased to be citizens in any true sense of the world' (Marshall 1950: 24). For Marshall (1950: 26), compulsory education instituted by the end of the nineteenth century was an indicator of the increasing attention to social rights. He discussed the state's difficulties to follow its obligation to guarantee benefits in the form of services (instead of cash) because of the qualitative element in the fields of health or education (Marshall 1950: 58). By regarding housing, Marshall stated:

> [T]he basic right of an individual citizen to have a dwelling is at minimal. He can claim no more than a roof over his head [. . .]. Nevertheless, the general obligation of the State towards society collectively with regard to housing is one of the heaviest it has to bear. (Marshall 1950: 60)

The high degree of interest (e.g., Isin & Turner 2002; the Journal 'Citizenship Studies'; Somers 2008) in Marshall's concept was, indeed, not without critical points. The following points are quoted from Turner (1990) who summarized critiques made by other scholars during the 1970s and the 1980s (for example by Giddens, Titmuss, Parsons). Turner added some of his own critical arguments. In this vein, Marshall failed:

- to consider a wider social context within which welfare policy developed in Britain;
- to perceive that additional social rights might be developed around culture;

- to explain how resources are to be generated and subsequently redistributed by the state to claimants, in terms of health provision and general welfare institution; and
- to mention that citizenship also emerged because the state tried to be a stabilizer of a social system as an answer to violence and threats of violence.

He was wrong when:

- he regarded citizenship rights as a unified homogenous set of social arrangements and
- he perceived the historical emergence of citizenship as an irreversible process within contemporary societies (Turner 1990: 192–194).

Another critical argument pointed out that Marshall was only regarding the rights of English men (Turner 1990: 195–196). Despite the above enumerated critical points, many aspects have changed since 1950, particularly the role and influence of the state (Turner 1990: 195). Citizenship rights, in their earlier understanding, were connected to the members of a community (e.g., a state) and excluded those who were not part of the community. In contrast, human rights always were (and still are) universal (U. Davy 2015). Every human being enjoys them inherently (Somers 2008: 6–7).

Since Marshall's essay, several scholars in economics became prominent due to their theories and models that explain in detail how and why wealth and inequality could grow side by side over the last few centuries (e.g., Deaton 2013; Piketty 2014). Social human rights are currently regarded more frequently in the context of globalization (Leisering et al. 2015). Deacon (2007: 3) stated that social policy should no longer be understood or made without reference to the global context. In Deacon's view, social policy since the 1980s has been undergoing a process of globalization (globalization of social policy); at the same time, global politics are characterized by a process of socialization (socialization of global politics) (Deacon 2007: 3). Globalization, following Deacon (2007: 9–10), has at least five impacts on the subject area and practice of social policy. It sets welfare states in competition with each other, brings new actors into the creation of social policy, raises the issues with which social policy is concerned to a supranational level, creates a global private market in social provision, and encourages people's global migration. Deacon's study revealed that the process of social policy-making is 'multi-sited, multi-layered, [and] multi-actored' (Deacon 2007: 175). This has consequences for researchers:

> Certainly, to make sense of even part of this complex reality, social policy analysts who have always been multidisciplinary animals gleaning (and rejecting) insight from politics, economics and sociology need now to engage with the theories of international relations and deploy an arsenal of research methods appropriate to international diplomacy, business forcasting [sic], the anthropology of the powerful and network analysis and probably much more besides. (Deacon 2007: 177)

Deacon (2007: 24–87) analyzed social policy advice global actors like the World Bank, IMF, WTO, OECD, ILO, the UN and its social agencies, as well as international non-state actors, have given to countries. Powerful states, powerful organizations, and even powerful disciplines (the economy, for instance) are 'in a war of position regarding the content of global policy' (Deacon 2007: 16). The debate about the land titling approach with many critics taking a stance against de Soto and the World Bank (4.2) will serve an example of this presumed war of positions regarding the contents of land policy as well. Kenna stated that globalization had a significant impact on housing systems:

> The growth and power of globalized corporations involved in housing; the globalization of property rights, housing finance markets, and the promotion of owner-occupation; globalized real estate investment in housing; the reordering of cities and slums; new roles for the state in relation to housing; and the effect of globalized migrants and refugees all present new and undetermined challenges. (Kenna 2008: 407)

Deacon's theories about the globalization of social policy as well as the socialization of global politics inspired academics to ask if there is such a thing as global social citizenship and what it could mean (B. Davy et al.

2013). The authors of a Supplement of the International Journal of Social Welfare (2013, edited by the principle investigators of FLOOR Benjamin Davy, Ulrike Davy, and Lutz Leisering) showed a somewhat broad understanding of global social citizenship. They presented different concepts to substantiate global social citizenship as a research category. Jacobs (2013) regarded the multifaceted character of private property with respect to social citizenship and human rights. He concluded that '[t]he challenge for the twenty-first century is to foster forms of private property whose benefits will be realized by those most in need of them' (Jacobs 2013: S99). U. Davy (2013 and 2014) regarded human rights as the international layer of social citizenship and identified three ingredients of social citizenship in the reports of the ICESCR member states: Individual welfare (targeting), inequality (poverty), and measures providing help where markets fail to do so (U. Davy 2014: 263). She proved that around 1993, the understanding of social rights in the ICESCR discourse changed:

> At that point in time, socialism and developmental thinking were in retreat. Liberalism was on the rise in the Eastern European transformation states, but also in Asian and African states. The reports testify to a new ideational consensus, cross-cutting former ideological groups and shared by states around the world. (U. Davy 2014: 263)

Leisering and Barrientos (2013) explored the role of social assistance and social cash transfers for the global poor in countries from the global South and North. They followed the widespread understanding that social citizenship is not only about having (receiving welfare benefits), but also about doing—what they understood as participation in life. Leisering and Barrientos (2013: S52) developed citizenship in three dimensions: resources; participation in common activities in markets, politics and civil society; and recognition, conveying a sense of belonging, dignity, and personal worth. B. Davy and Pellissery (2013) considered the human right to housing as a promise of social citizenship and claimed that its implementation in informal settlements is not possible without full consideration of the spatiality and sociality of non-state welfare. Dean (2013) stated that social rights are constructed through the naming and claiming of needs. He stressed the multidimensional character of social citizenship and pointed out that '[h]uman needs are defined and redefined locally and globally, across time and space and through lived experience' (Dean 2013: S36).

Social citizenship puts the individual, including his or her social rights, into the center of attention. In the case of housing, taking such a perspective helps to regard people living in informal (or illegal) housing not as rights violators, but as right holders. Furthermore, as housing is a complex issue, social citizenship often stands for claiming individual solutions (targeting, for instance) instead of large-scale measures (such as land reforms). Human rights play a significant role because they 'contribute to global social citizenship not only as a body of law and a monitoring mechanism but also as a discourse that mobilizes a global public' (B. Davy et al. 2013: S10).

For the identification of social citizenship in the ICESCR discourse, three elements have been explored: rights, inclusion, and equality. I suggest not talking about social citizenship states, but instead—picking up the metaphor of B. Davy et al. (2013)—asking if and to what extent the ICESCR member states entered the road to social citizenship and, if they did, if they even entered the road to global social citizenship. States have entered the road to social citizenship when they reported the following three elements with a remarkable amount of attention in their States Parties reports.

Rights. States that entered the road to social citizenship acknowledged that people affected by inadequate housing were, to some extent, able to help themselves. A state entered the road to social citizenship when the state regarded people affected by inadequate housing as individuals—as right holders or as entrepreneurs, for instance—instead of as an anonymous mass or as rights violators. The mentioned measures often concentrated more on the individual, for example, when the state talked about microcredits, land titling, targeted financial assistance, or targeted health care instead of large-scale programs such as mass construction programs. States entered the road to social citizenship when they stopped referring to the people affected by inadequate housing as criminals who had to be evicted. Instead, they spoke about informal settlements rather than illegal settlements, or about protection rights (protection from eviction or expropriation, or compensational measures).

Inclusion. States on the road to social citizenship usually paid sufficient attention to the complexity of inadequate housing. They incorporated vulnerable and disadvantaged groups when they talked about housing. I examined the mention of four groups in detail: refugees, asylum seekers (refugees who make an application for the right to stay), indigenous people, and Roma.[1] By reporting these and other disadvantaged and vulnerable groups, the states confirmed their recognition of both the heterogeneity and the complexity of inadequate housing.

Equality. States on the road to social citizenship discussed housing policies under the umbrella of equality as a policy goal. They explicitly mentioned equality goals, mostly by describing inequality as a problem that they wanted to solve. Regarding housing, this often means that they showed a self-critical view by acknowledging how difficult it was to reach equality in the context of housing. Equality certainly does not mean that every person should live under the same housing conditions, but rather an equality of status and rights for those affected by inadequate housing. Equality as a policy goal could mean, among other things, that a homeless person or a refugee should also be able to claim their rights to health or social security.

For this book, global has two meanings. First, the analysis is global because almost all countries worldwide have ratified the ICESCR (even though the United States did not) and most of them have submitted at least one State Party report. The findings are based on the States Parties reports of 132 countries worldwide. Second, a social policy on a global level is characterized by three indicators. It (1) incorporates supranational actors (also called global actors), (2) responds to global migration, and (3) points out the significance of international cooperation by responding to social human rights. States that reported the three elements of social citizenship, as well as showed the three indicators of

[1] Drawing from the 'Descriptive Glossary of terms relating to Roma issues' published by the Council of Europe (2012), I use the term Roma when the states or the Committee reported Roma, Sinti, or spoke of 'Gypsies.'

global social policy, are the states that entered the road to global social citizenship.

The many needs of people affected by inadequate housing can be defined globally or on a national level, but de facto, people will meet their needs on a local level. The state is not the only responsible actor, anymore, when it comes to the implementation of the human right to housing; many other actors on global as well as on local levels are involved.

> On the one hand, globalization has diminished the ability of all states to control economic outcomes that affect the well-being of citizens. [. . .] On the other hand, states remain the central actors in economic planning even if they are often pounded by [. . .] forces of economic globalisation. Although their economic power has diminished they still can enact policies to respect, protect and fulfil economic and social rights. (Felice 1999: 585–586)

One central question in both social policy as well as housing research is: 'Should the state, market, organisations of civil society, the family and kin provide for welfare needs of the population, and in what proportion?' (Deacon 2007: 5). Gough and Wood (2004) (and the other authors who published in this collection of essays) developed a welfare regime approach in which they applied the welfare debate to global regions that have not been on the agenda of many Western scholars as much. Their model stresses the importance of informal arrangements and helps scholars to think outside of their Western point of view by understanding welfare policies in the global South. They understood social policy as 'a public policy [. . .], oriented to social welfare goals [. . .], which operates through a wide variety of policy instruments across a number of sectors [. . .] and is formulated and implemented by a wide range of actors' (Gough 2004a: 22). In contrast to Esping-Andersen (2011, 4.3), who observed the welfare regimes in 18 OECD countries, the research group was interested in countries in the developing and poor world. In these countries, neither a legitimate state exists nor a pervasive, formal sector labor market, as Wood (2004: 83) stated. Therefore, the researchers used the term welfare regime instead of welfare state regime:

> Welfare state regimes refer to the family of social arrangements and welfare outcomes found in the OECD world of welfare states. Welfare regime is a more generic term, referring to the entire set of institutional arrangements, policies and practices affecting welfare outcome and stratification effects in diverse social and cultural contexts. Thus welfare state regimes form one 'family' of welfare regimes alongside others. (Gough 2004a: 26–27)

In opposition to a welfare state regime, where people can expect to meet their security needs via participation in labor markets, financial markets, and the finance and provision role of the state, informal security regimes emphasize the role of community and family relationships (Gough 2004a: 33). In contrast to informal security regimes, in welfare state regimes, the dominant mode of production is capitalism, the social relationship is dependent on market inequalities, and the source of livelihood is the labor market. The situation in informal security countries is more complex and confusing, particularly regarding the institutional responsibility matrix (Table 4.1). Gough (2004a: 30) added the actor group of community to the classical state-market-family trinity (Esping-Andersen 2011). Under this term, he summarized informal groups (civil society) as well as formal NGOs. He explained that, in informal welfare regimes, a wider range of institutions and actors is involved and, in contrast to the welfare state regime, all four elements have important counterparts at the supranational level. In this case, Gough (2004a) agreed with Deacon (2007: 175) who also emphasized the global nature of the social policy making process.

Table 4.1 Components of the institutional responsibility matrix

	Domestic	Supra-national
State	Domestic governance	International organizations, national donors
Market	Domestic markets	Global markets, multi-national cooperation
Community	Civil society, NGOs	International NGOs
Household	Households	International household strategies

Source: Gough 2004a: 30

Gough (2004a: 30) conceded that the institutions (the actors) do not operate independently of each other. Self-interest is not confined to the market realm, nor loyalty to the family realm, neither group interests to the political realm. Instead, there is a sort of permeability. Behavior is often the same when acting within the state, the market, the community, or the family (Gough 2004a: 31).

Gough (2004a: 34) referred to other articles of the book where the regime genuses and species in countries from East Asia (Gough 2004b), Africa (Bevan 2004), and Latin America (Barrientos 2004) were studied. On a global level, the most used indicators for measuring and differentiating between welfare systems are often welfare outcomes, such as the Human Development Index (HDI), disability-adjusted life expectancy (DALE), the index of adult literacy, or the World Bank's poverty gap ($2 a day). However, welfare mixes also often describe the relationship between the expenditures of the four components of the institutional responsibility matrix (Gough 2004a: 37–39). Gough (2004a: 39–49) admitted that the welfare mix cannot be operationalized in its full scope due to a lack of available data for the expenditures of the community and household sector. The responsibilities are, however, a useful concept to examine in terms of how social policies differ from each other in a global context. Again, the role of the state is in the center of attention: '[S]ocial policy of poor countries very quickly moves onto the agenda of "civil society compensating for inequities of the state" instead of the OECD welfare state regime principle of "the state compensating for inequities of the market"' (Wood 2004: 84).

As the present analysis aims to be a global one, seven groups of actors were generated that should fit with the complexity of responsibilities in all considered States Parties reports (Table 4.2). I aimed to capture actors from the local to the global level as well as actors whose obligations and interests are usually different by considering the implementation of the human right to housing.

Governmental actors have by means of law the obligation to guarantee adequate housing. However, they can respond to the right to housing in different ways, meaning that they are not obliged to build homes for everyone (UN-Habitat & OHCHR 2002: 17). Usually, their organizational structure is hierarchical. As the

Table 4.2 Actors responding to inadequate housing

Actors	Differentiation	Examples
Governmental actors	Formal obligation to respond to inadequate housing	The states, governments, ministries, regional planning
Local actors	Formal obligation to respond to inadequate housing on a local level	Local authorities, municipalities, burgomasters
Family/self-help	Formal or moral obligation to help relatives or themselves in the case of inadequate housing	Family and kin, self-help, parents, individual initiative, households
Market actors	Interest to maximize profit	Private building sector, banks, business associations, developers
NGO+	Interest or moral obligation to respond to inadequate housing	Voluntary sector, non-profit organizations, churches, religious institutions, charitable associations, civil society
Supranational actors	Interest and/or obligation to respond to inadequate housing in more than one country, usually in many countries worldwide	The Cities Alliance, World Bank, other states, ILO, UNDP, the EU
Other actors	Different obligations and interests with respect to inadequate housing	Experts, media, working groups, round tables, networks, task force

ICESCR States Parties must report their policies responding to the human right to housing, I expected governmental actors to dominate in the States Parties reports. However, many different governmental actors exist in each member state. The distribution of responsibilities between national, regional, and local levels depends on each country. As Dean stated, 'many needs are defined and redefined locally and globally' (Dean 2013: S36; see above). Housing always has a spatial dimension. In the end, the human right to housing can only be met locally, even if the mentioned housing policy or program is described as a national one. For this reason, I generated a second group that stands for governmental actors on the local level. I call this group **local actors**.

Due to **family and self-help** playing an important role in social policy discourses (Esping-Andersen 2011; Gough & Wood 2004) they are also examined as responsible actors in the States Parties reports. The terms family or household came up frequently in the ICESCR discourse, such as when a State Party talked about family rights, or when households were reported as the recipients of financial assistance. Of course, in such cases, family or self-help did not count as actors.

Market actors usually have an interest to maximize their profit, or, at least, to avoid debt. They usually have no direct obligation to respond to inadequate housing or to guarantee adequate housing, unless such an action would be profitable. Their incidence was also regarded as an indicator to testify how significant the economic factor on the road to social citizenship in the ICESCR discourse really was.

The term to summarize domestic actors that are neither governmental nor market actors is **NGO+** (spoken: NGO plus). NGO+ includes different NGOs, non-profit organizations, charitable associations, and religious institutions, among others. Their basic motivation to deal with inadequate housing does not usually fulfill a legal obligation, but is a voluntary act of solidarity.

With respect to their interests, and formal and moral obligations, the group of **supranational actors** is heterogeneous. Supranational actors respond to inadequate housing in a global, or at least in a supranational context. The group includes governmental actors (such as when a State Party reports the support of another state), INGOs, global players such as ILO, the World Bank, UN-Habitat, UNICEF, or complex bodies such as the European Union. In contrast to Gough (2004a: 30; Table 4.1, p. 87), I did not differentiate whether a mentioned supranational actor was a governmental actor, a market actor, or an INGO.

The group of **other actors** is also heterogeneous, but for other reasons. It is comprised of actors that have no obligation, but rather a seemingly coincidental interest to respond to the human right to housing. I also coded the group of other actors when a State Party reported roundtables, working groups, networks, or when a mentioned actor could not be identified as one of the other six groups of actors.

Of course, the actors do not always perform as the differentiation in the table indicates. Governmental actors are complex, and they often act as market actors, for example, in the field of social housing. Housing cooperatives, for example, are also often market actors, but have their roots in solidarity and voluntarism. However, examining the actors in the States Parties reports is the key to lots of subjects that social policy research is interested in, such as the changing role of the state in the discourse narrative. Moreover, the incidence of supra-national actors is a key indicator to testify how global a State Party's consideration of the human right to housing is.

This section has introduced social citizenship and initiated global social citizenship as a research category for the ICESCR discourse. Moreover, certain actor groups have been developed to answer the responsibility question. The next section points out the significance of land for people affected by inadequate housing.

4.2 Land Policy and Inadequate Housing

The availability of land is the first essential condition for a well-functioning housing sector (Angel 2000: 192). This does not mean, however, that the term land comes up very often in housing discourses. Frequently, land is hidden in terms like property, real estate, location, settlement, place, tenure, space, and many more. As a research discipline, land policy examines the allocation of land rights and the distribution of the gains and losses of land uses (B. Davy 2005) and is, therefore, essential for planning decisions (Oxley 2004; Ploeger & Groetelaers 2007). Land policy scholars observe housing (Alterman 2002; B. Davy & Pellissery 2013; Kolocek 2012 and 2013), flood plains (Hartmann 2011), poverty (B. Davy 2009; Deininger 2003), dignity (B. Davy 2014), or public spaces (Webster 2007), to name a few examples. Property in land can include various rights like the right to enjoy and use, the right to transfer, develop and improve, or the right to benefit from increased property values or rental income (UN-Habitat & GLTN 2008: 6). UN-Habitat and the Global Land Tool Network (GLTN) defined property rights as '[r]ecognised interests in land or property vested in an individual or group' (UN-Habitat & GLTN 2008).

Ostrom pointed out that at least two individuals are necessary for a working property rights relationship:

> Property rights define actions that individuals can take in relation to other individuals regarding some 'thing'. If one individual has a right, someone else has a commensurate duty to observe that right. Squatters do not possess property rights even though they may be the users. (Ostrom 2001: 134)

Fifty years earlier, T.H. Marshall stated that '[a] property right is not a right to possess property, but a right to acquire it, if you can, and to protect it, if you can get it' (Marshall:1950: 35).

The human right to property (Article 17 UDHR) and the human right to an adequate standard of living (Article 25 UDHR) are of essential significance for housing and for land policy (B. Davy 2012: 167–169). Article 17 UDHR can both strengthen and hinder land policy goals. Jacobs (2013: S96) stated that (private) property rights can be essential for the implementation of many human rights, but they can also impede their realization. By considering Article 25 UDHR and Article 11 ICESCR, land policy stresses that an adequate standard of living requires at least minimal access to land:

> [N]o person can exist without using land. Everybody occupies physical space, but the full participation in everyday life also requires that land be available for housing, work, mobility, leisure, and social exchange. The scope and content of the right to minimal access to land may vary and cannot be expressed in square meters or requirements for land rights formalization. Minimal access to land comprises the individual right to all land uses which are indispensable for a person to achieve an adequate standard of living. (B. Davy 2012: 170)

The increase in land values was for a long time regarded as the main cause of inadequate housing conditions and expulsion:

> The growth of the big modern cities gives the land in certain areas, particularly in those which are centrally situated, an artificial and often

> colossally increasing value; the buildings erected on these areas depress this value, instead of increasing it, because they no longer correspond to the changed circumstances. They are pulled down and replaced by others. This takes place above all with workers' houses which are situated centrally and whose rents, even with the greatest overcrowding, can never, or only very slowly, increase above a certain maximum. They are pulled down and in their stead shops, warehouses and public buildings are erected. Through its Haussmann in Paris, Bonapartism exploited this tendency tremendously for swindling and private enrichment. But the spirit of Haussmann has also been abroad in London, Manchester and Liverpool, and seems to feel itself just as much at home in Berlin and Vienna. The result is that the workers are forced out of the centre of the towns towards the outskirts; that workers' dwellings, and small dwellings in general, become rare and expensive and often altogether unobtainable, for under these circumstances the building industry, which is offered a much better field for speculation by more expensive houses, builds workers' dwellings only by way of exception. (Engels 1942 [1872]: 19)

The combination of land policy and social policy is not new. Instead, it has a long tradition. More than 100 years ago, land reforms were prominent instruments of land policy to obtain social justice (Damaschke 1918 [1902]; B. Davy 2000). Paine (1903 [1797]) questioned in the late eighteenth century the inviolability of private property and claimed a reform that obligated landowners to pay into a national fund out of which 15 Pound Sterling per annum should be paid to every person older than 20 years and 10 Pound Sterling per annum to every person age 50 or older (Paine 1903: 400; see also B. Davy 2012: 29–30; Marangos 2008: 317). Paine's contribution combines land policy and social policy because 'Paine anticipates the significance of property in land for social citizenship' (B. Davy 2012: 29). In the early twentieth century, Damaschke described the housing situation of working-class families living in Berlin. Fast growing land and rent prices led to overcrowded and unhygienic housing situations. Damaschke demanded a land reform to distribute the gains of industrialization to all people, not just the landowners: 'The housing question is particularly a land question' (Damaschke 1918: 87, author's translation). As B. Davy (2012: 29) concluded, the success of the European welfare state made it easy for many political parties to drop land reform, and now land reform is more often associated with expropriation or confiscation than with the mitigation of social problems. Nevertheless, the spatial consequences of property (B. Davy 2012) can be either the cause or the solution for people living in situations of inadequate housing.

As a social construction, land has an exchange value, an existence value, a use value, and a territorial value (B. Davy 2012: 89–136). Land can be understood as a commodity, but also as capability, territory, and environment (B. Davy 2012: 91). As a commodity, land differs from other goods. The commodification of land is a precondition to private property: 'In order to be owned, land has to be commodified and it has to be sufficiently clear where a plot of land starts and where it stops' (B. Davy 2012: 113). Policymakers can commodify land through land law and land use planning (B. Davy 2012: 113–114). The best-known example of the commodification of land is probably land titling.

Land titling means 'the allocation of real property rights on land, e.g., rights that are opposable to a third party, and that can be transferred, inherited and mortgaged' (Payne et al. 2009: 444). Land titling became prominent through Hernando de Soto (2000) and the World Bank (Deininger 2003; Deininger & Binswanger 2001). Based on different global experiences, de Soto (2000) explained why capitalism triumphed in Western countries and failed everywhere else: 'Entrepreneurship triumphed in the West because the Law integrated everyone under one system of property, giving them the means to cooperate and produce large amounts of surplus value in an expanded market' (de Soto 2000: 71). The author had both informal working and informal housing in mind. He emphasized the significance of property rights by answering questions about wealth and poverty. Perhaps the property-wealth causality sounds trivial, but de Soto did not simply regard 'property [as a] physical thing that can be photographed or mapped. Property is not a primary quality of assets but the legal expression of an economically meaningful consensus about assets' (de Soto 2000: 167).

So far, the key message—wealth for the poor through property rights—sounds simple, but the procedure is not. De Soto (2000) called the procedure capitalization process. He listed four successive strategies that are each subdivided into many smaller successive steps (de Soto

2000: 160–161). In the following, I will present the most important steps and aspects to give an impression of the complexity of the formalization process.

The discovery strategy is separated into five parts, starting with identifying, locating, and classifying the extralegal assets (dead capital). Next, the actual and potential value of extralegal assets must be quantified, the interaction of the extralegal sector with the rest of society must be analyzed, the extralegal norms that govern extralegal property must be identified, and the costs of extralegality to the country must be determined. All the listed steps are separated again into smaller steps.

The political and legal strategy is divided into six basic steps. The first two steps ensure that people at the highest political level assume responsibility for the poor and put the appropriate agencies into operation. After this, administrative and legal bottlenecks must be removed, and a consensus between legal and extralegal sectors must be built. Next, the costs of legally holding assets must be reduced in comparison to the costs of holding assets extralegally. The last step is to create mechanisms that will reduce risks associated with private investment.

The operational strategy has three basic steps, but is the most complex due to many smaller steps. Field operation strategies, personal equipment, and offices must be designed and implemented. Communication and participation is needed, as well as computer knowledge, for the registration process.

The last strategy is the commercial strategy. This strategy deals with the implementation and enforcement of credit systems, insurance products, or infrastructure (de Soto 2000: 160–161).

When considering de Soto's capitalization process, many overlaps to other planning processes become visible: identification of potentials, examination of the bordered area, information, the participation of different actors, mapping, and monitoring. Although de Soto (2000: 160–161) took only two pages to describe the capitalization process, planners will recognize that most of the mentioned single steps—even the small ones—take a considerable amount of time and financial resources. That is not, however, the main reason why scholars criticized de Soto's approach (Campbell 2013; Davis 2006; Gilbert 2002; Neuwirth 2011; Payne 2001; Payne et al. 2009; Roy 2005 and 2010). The key critical argument did not deal with the complexity, but with its underlying philosophy. Land titling was often regarded and criticized as a Western, neo-liberal approach that would not work in countries of the global South. The rise of neo-liberalism was often associated with negative impacts on the poor (UN-Habitat 2003: 3), particularly with rising inequality (Pogge 2007). Other scholars questioned if the poor would really be the ones to profit:

> As such, many argue that massive titling programmes are being conducted for reasons that have nothing to do with helping the poor. Such programmes are popular because they are cheap; it is much less expensive to issue property titles than to provide settlements with services. In addition, the authorities and international agencies can actually make money from titling programmes. The World Bank has long recognised that the profits made by a government agency may be used to finance upgrading programmes elsewhere [...]. Indeed, many people in Washington argue that issuing legal titles on a large scale can only be justified if the beneficiaries are prepared to pay the full cost of the process. (Gilbert 2002: 7)

We find a similar argument from Davis:

> Ironically, de Soto, the Messiah of people's capitalism, proposes little more in practice than what the Latin American Left or the Communist Party of India (Marxist) in Kolkata had fought long for: security of tenure for informal settlers. But titling [...] is a double edged sword. Titling [...] accelerates social differentiation in the slum and does nothing to aid renters, the actual majority of the poor in many cities.[...] However, de Sotoan approach panaceas remain immensely popular for obvious reasons: the titling strategy promises big social gains with a mere act of the pen and, thus, pumps life back into World Bank's tired self-help paradigms; it accords perfectly with dominant neoliberal, anti-state ideology, including the Bank's current emphasis on governmental facilitation of private housing markets and the promotion of broad home ownership; and it is equally attractive to governments because it promises them something—stability, votes, and taxes—for virtually nothing. (Davis 2006: 80–81)

Of course, tenure security is a fundamental component of the right to housing (UN-Habitat & OHCHR 2002: 42). Property rights are associated with tenure security, but, in fact, this is not the same. Land tenure security depends on different aspects like the confidence that land users will not be arbitrarily deprived of the rights they enjoy, the certainty of recognition by others, or the protection from eviction (UN-Habitat & GLTN 2008: 6). UN-Habitat and GLTN listed 11 tenure systems with different characteristics, advantages, and disadvantages (UN-Habitat & GLTN 2008: 9–10). Having a property right in land does not mean being protected from arbitrary eviction. Payne et al. (2009: 447) mentioned that titling could even cause the reduction of tenure security. Van Gelder (2013) claimed that a state that supports land titling (of formerly illegally occupied land) paradoxically sends out the signal that illegal occupation, which might be a violation of property rights, could be understood as the first step to be integrated into the formal system of property rights.

The World Bank (hereafter also: the Bank) was regarded as De Soto's main partner by supporting land titling (e.g., Deininger 2003; Deininger & Binswanger 2001; Gilbert 2002; Jacobs 2013: S94–S96). The Bank is here presented in a short excursion because the Bank is the most prominent global actor that connected social policy with land policy. The history of the World Bank is also a history of housing, land policy, and global social policy.

The World Bank was created in 1944 at the Bretton Woods conference 'as an institution for funneling capital to the war-torn global periphery in the face of liquidity-constrained capital markets' (Gilbert & Vines 2000: 42). The other two Bretton Woods Organizations are the International Monetary Fund (IMF) and the World Trade Organization (WTO, since 1994, formerly GATT) (Deacon 2007: 142). Formally, the World Bank is an agency of the UN family, but the World Bank differs in many aspects from the UN, such as in voting rights (Deacon 2007: 143) or with respect to its responsibilities associated with human rights (Oestreich 2004).

After World War II, the World Bank's focus was first on reconstruction. Its policies were dominated by 'the intensification of the Cold War and the need to bolster the democratic world against the Soviet threat' (Gilbert & Vines 2000: 14). As its reconstruction role in Europe became redundant by the late 1950s, the Bank concentrated more on underdeveloped countries, particularly in Africa. The priorities shifted from reconstruction to development. During the 1970s, the Bank increasingly concentrated on education and health and spatial topics like rural development and urban infrastructure. Urbanization and the increasing squatter settlements led the Bank to extend its range of developmental loans to infrastructure and housing (Pugh 1995: 64).

In 1980, the first structural adjustment loan was launched (Ferreira & Keely 2000: 159), and the era of structural adjustment policies (programs and reforms) started. Structural adjustment policies resulted from the oil shocks and the debt crisis in the 1970s (Ferreira & Keely 2000: 161). The goals of Structural Adjustment Programs (SAPs) were to improve the standard of living through intermediate targets like ensuring a higher level of economic growth and maintaining its stability (Noorbakhsh & Noorbakhsh 2006). While stabilization policies dealt with the demand site and involved the two components of expenditure reduction and expenditure switching, adjustment referred to reforms of the supply side (Ferreira & Keely 2000: 160–167). Structural adjustment programs should support the policy reform in the fields of fiscal, monetary and exchange policy, or privatization, trade liberalization, and financial sector reform (Deacon 2007: 27). SAPs have been criticized intensively. Noorbakhsh and Noorbakhsh (2006) summarized the critiques. The main points revealed that SAPs had adverse effects on the standard of living in poor countries, sometimes led to misallocation of scarce resources or wasted public funds, affected the aid provided by donor countries, proved inappropriate for African countries, and raised inequalities (Noorbakhsh & Noorbakhsh 2006).

Despite all criticism, structural adjustment has encompassed a large range of reforms in the sectors of trade, price, tax or labor market; in the financial sector; and in the fields of privatization and institutional building (Ferreira & Keely 2000: 166–167). The Bank's philosophy during the 1990s is known under the term Washington Consensus. The term was introduced by the economist John Williamson and 'is generally interpreted as referring to the neo-liberal belief that the combination of democratic government, free markets, a dominant private sector and openness to trade is the recipe for the prosperity and growth' (Gilbert & Vines 2000: 16).

By regarding the critical arguments, one must consider that the Bank still is and always was in a situation that Oestreich described as the 'too much/too little conundrum' (Oestreich 2004: 57). The author observed the Bank's responsibilities with respect to human rights and stated that the Bank has, 'despite its affiliation with the UN, [. . .] at best only a loose commitment to promote human rights and other values beyond economic development' because '[t]o pursue values other than economic efficiency, it must show some sort of mandate' (Oestreich 2004: 58), which it does not have.

> To some extent, then, the World Bank is taking steps toward doing what ought to be expected of it, albeit belatedly; that is, it is attempting to reform its operations to fit with generally accepted ethical principles while not overstepping its limited authority to be an agent of social change. Those who would like to see the institution take a more wide-ranging approach to human rights promotion ignore the institution's limitations both in its authority and its financial capabilities; those who would like to see the institution act in a completely apolitical manner ignore the fact that it operates only at the sufferance of international civil society. (Oestreich 2004: 69)

Pugh (1995: 39–40) noted that the Bank's non-housing policy in the 1980s—loan financing for governments, deregulating markets, ensuring private property rights, achieving macroeconomic stability, enhancing of the development of finance capital markets—had relevance for housing and poverty. Even when this development occurred in the fields of moderate or median income groups, the poor were affected indirectly, because they were at the end of this housing system (Pugh 1995: 43–44). By regarding outcomes of the stabilizations and structural adjustment programs in Costa Rica, Kenya, Mexico, and Brazil, Pugh (1995: 43–54) concluded that macroeconomic factors like inflation, recession, increasing rates of interest and curbs in public spending for infrastructure had an impact on housing. The poor had been affected both by direct income and housing-related consequences and by direct impacts such as reduced housing supplies.

By describing the situation from 1972 to 1992, Pugh (1995) differentiated between three phases of the Bank's housing policy. In the first phase, neo-liberalism with the emphasis of individualism, free markets, and user pay principles dominated economic policy (Pugh 1995: 66). This policy contrasted the subsidized public housing approach. The challenge was to make housing affordable for low-income households without the payment of subsidies: 'Cost-recovery would reinforce affordability' (Pugh 1995: 64). In this phase, lots of site-and-services projects and complementary slum upgrading schemes were implemented. The states were not responsible for creating housing credit systems and reforming property rights; the state's task was only to install infrastructure and provide tenure rights for occupiers (Pugh 1995: 65).

In the second phase, beginning in the early 1980s, housing credit institutions spread because housing was related to developments on the macroeconomic level. The World Bank adopted the policies of the United States Agency for International Development (USAID). Countries such as India and Sri Lanka received intermediate loans for housing sector development (Pugh 1995: 66). In many developing countries, however, the World Bank found economic problems and inadequacies. 'In short, in some countries reform was required, and in others initiatives were necessary to create new housing finance institutions' (Pugh 1995: 66). Pugh (1995: 66–67) saw various implications for housing policy in this second phase. First, the nature of the reforms of the third phase was indicated; second, the approach led to a more rapid disbursement of loan funds; third, the poorest households were often bypassed by the formal housing finance institutions; fourth, the new policy concentrated on the housing system as a whole and finally, the first phase provisions of site-and-service and slum upgrading approaches were not replaced, but rather encompassed in a wider policy framework. The problem with the second phase was its self-limiting factor: It was confined to finance, which omitted land policies, the construction industry, infrastructure, and deregulation in town planning (Pugh 1995: 71).

Between 1986 and 1992 (third phase), the main thrust was the growth and development of the entire housing sector in its urban and national context (Pugh 1995: 67). Globalization came to be regarded from a more local perspective, which complicated the situation. In 1986, the United Nations Center for Human Settlement (UNCHS), the

United Nations Development Program, and the World Bank collaborated in founding the Urban Management Program (Pugh 1995: 67). Their aim was to improve their performance in developing countries in land management, municipal finance, infrastructure services, the environment, and to build up the capacity of urban management institutions. Housing should be linked to the wider urban economy (Pugh 1995: 67). The new philosophy was enablement, which was defined 'as providing the legislative, institutional, and financial framework whereby entrepreneurship in the private sector, communities, and among individuals can effectively develop the urban housing sector' (Pugh 1995: 67–68). Enablement was a part of the new political economy (NPE), which afforded more attention to the governments' roles than simpler and earlier neoliberalism had done. In contrast to the first phase, the new enablement required partnerships and interdependence among state agencies, markets, NGOs, and individuals' self-help (Pugh 1995: 68). Particularly in the field of land management, the state's role remained significant, for example, with respect to property rights reforms or land registration systems. In 1991, the World Bank again increased the attention given to urban policy. Pugh (1995: 36) called the report Urban Policy and Economic Development (World Bank 1991), a forerunner of the strategic review of housing policy. The new housing policy should connect urban growth to macroeconomic policy and more: 'Urban policy was to be formulated at national levels to be connected with macroeconomic conditions, especially in reforming housing finance institutions and legislations on land development and town planning' (Pugh 1995: 69). Pugh's summary of the third phase as a 'package of reforms' (Pugh 1995: 70) underlines the complexity of housing policy and moreover shows a tendency toward self-help.

At present, the Bank plays an important role by reaching the Millennium Development Goals (MDGs). The Bank works in partnerships with numerous NGOs (Oestreich 2004: 70) and as a member of Cities without Slums. Each year, the World Bank publishes World Development Reports which are sometimes quoted in the States Parties reports. The view on titling approaches has changed, too. The proponents of land titling started to acknowledge that a simple big scale land titling approach would not succeed very often. The land reform experiences in Latin America (de Janvry et al. 2001) and many other experiences from small-scale projects all over the world demonstrated that the special circumstances of the location had to be answered with flexibility. Meanwhile, the World Bank recognized that more nuanced approaches are desirable (Payne et al. 2009: 445).

The land titling approach as well as the other World Bank policies for poverty alleviation can be understood here as a form of market-orientated, or liberal, social policy that responds to informality.

Informality, however, is difficult to define (Fernandes 2011: 10–12). Scholars use different terms for informality, such as 'System D' (Neuwirth 2011), 'black economy' (Castles and Ferrera 1996: 178), or 'shadow economy' (Schneider & Buehn 2009). In simple terms, informality means the absence of governmental planning and legal regulation (Roy 2005: 149). It steps up when planning fails, becomes overtaxed, or is ineffective (UN-Habitat 2009b: 134). When a state does not provide basic protection, people are forced to seek alternatives to meet their needs, and informal settlements could be regarded as a visual expression of such strategies (van Gelder 2013: 502–503). By taking these arguments seriously, the number of people living in informality could be regarded as an indicator of the absence or failure of welfare policy. Nonetheless, '[i]n practice, informal activities like formal activities, comply with rules, although the sources of rules and the means through which they are specified and enforced are different from laws governing formal activities' (UN-Habitat 2009b: 133–134). Sometimes, informality is regarded as 'second-class rights for second-class citizens' (Fernandes 2011: 5). De Soto pointed out that requirements to start a legal business are often one basic reason for informal—or, in his words, 'extralegal'—structures:

> [M]ost people in developing and former communist nations cannot get into the legal property system, such as it is, no matter how hard they try. Because they cannot insert their assets into the legal property system, they end up holding them extralegally. (de Soto 2000: 52)

Nevertheless, informal sectors are not separated, but characterized as 'a series of transactions that connect different economies and spaces to another' (Roy 2005: 148). Typical informal sectors are the informal housing sector and the informal working sector. Slums and homelessness are examples of informal housing. Typical examples of informal working are street vending, prostitution, windscreen washing at traffic lights, collecting waste or returnable bottles (Fragemann 2014; Moser 2014), and shoe shining. Informal housing and working sectors are sometimes interdependent and difficult to separate.

Roy (2005), de Soto (2000), and other scholars discussing informality, mostly had informal structures in developing countries in mind. Upon a closer look, however, many informal housing and working structures exist in Western countries, too. Begging and squatting are only the major examples. Children selling lemonade in front of their houses could be regarded as part of an informal economy (Neuwirth 2011: 150). The quite romantic idea of children who sell lemonade demonstrates impressively that the supposed 'absence of planning and regulation' (Roy 2005: 149; see above) is not at all bad in many cases. Thus, I propose to supplement the previous explanation of informality: Informality arises when planning fails, gets overtaxed, is ineffective, or when planning and regulation are simply unnecessary.

After the 'communicative turn' (Healey 1992) new forms of communication and participation planning began to spread. In Western countries, to a certain extent, informal aspects of planning became formalized. Nevertheless, informal planning still sounds like a contradiction of terms, but stands for different approaches on national, regional, and local planning levels. The German planning system, for instance, offers a variety of informal instruments (the German term is Informelle Planung) like national competitions, regional cooperation models, city networks, information, moderation, round tables, and other participation models on the local level (Gorsler 2002).

Informality is sometimes caused by formal legal decisions. After the implementation of the new German container deposit legislation in 2002 (VerpackV), a new informal economy spread through many German cities, wherein people collect and receive refunds on plastic bottles and cans (Moser 2014; Fragemann 2014). Some homeless people in Germany earn money by selling street magazines without paying taxes. However, scholars seldom discuss such examples under the umbrella of informality. In many European countries, people working or housing in informal structures have long been ignored by the governments as well as—at least to some extent—by scholars.

Informal housing in the global South, too, brings policymakers into difficulties. Van Gelder (2013) addressed a series of paradoxes that apply when governments try to deal with urban housing informality. One of them addresses the problem that governments often both penalize and simultaneously legitimize the act of informally occupying land (van Gelder 2013: 495). By examining 63 documents of the UN Special Rapporteur and Independent Experts on the right to adequate housing, B. Davy and Pellissery found that the reports presented the dwellers of informal settlements 'as victims who are vulnerable to environmental pollution and discrimination, are without secure titles and good services, are subjected of forced eviction or demolition' (B. Davy & Pellissery 2013: S76). The authors suggested looking at informality not as a problem of inadequate housing, but as a solution. By commenting on the reports of the UN Special Rapporteurs, they criticized that the 'reports hardly recognise how informality enables the poor to enjoy significant aspects of their human rights *because* they are living in informal settings' (B. Davy & Pellissery 2013: S77, emphasis in the original). The authors called the rules that underlie this alternative way the 'everyday social contract of informality' and stated that '[u]nless a government accommodates the everyday social contract of informality in its decisions and actions, it will not be able to protect, respect and fulfil the human rights of persons living in informal settings' (B. Davy & Pellissery 2013: S78). In contrast to de Soto's approach, the everyday social contract of informality is not a starting point for change toward formalization. Instead, the authors regarded it as a successful survival strategy that 'comprises rules based on trust (not obligation) that resist the pressure of formal institutions such as the rule of law or the formal economy' (B. Davy & Pellissery 2013: S79). A similar argument was made by Neuwirth:

> Security, stability, protection, and control are what's important. It doesn't matter whether you give people title deed or secure tenure, people simply

> need to know they won't be evicted. When they know they are secure, they build. They establish a market. They buy and sell. They rent. They create. They develop. Actual control, not legal control is the key. Give squatters security and they will develop the cities of tomorrow. (Neuwirth 2006: 302)

Scholars increasingly regard informality not only as a contradiction to rights, but also as a concept that contradicts illegality. The key difference between informality and illegality (as terms) is that informality often stands for something good while illegality often does not. Whereas informality means the absence of planning and regulation, I regard illegality as the violation of planning and regulation. To call a settlement illegal suggests that the speaker more likely regards its dwellers as criminals, as violators of rights. To call the same settlements informal implies that its dwellers are considered as victims, entrepreneurs, or citizens, albeit second-class citizens (Fernandes 2011: 5, see above)—but not criminals.

This section has first introduced the best-known land-based approach in the face of the social policy, namely land titling. The discourse on land titling has proven that the view on land as a commodity is often an issue of controversial discussion. I presented the World Bank as an influential actor in the face of land policy as social policy. Scholars, who criticized land titling, clamor for other informal arrangements that are based on trust, security, stability, protection, and control. In the ICESCR discourse, the relationship between the terms informal and illegal is a simple but useful indicator to explore how the states' consideration of people affected by inadequate housing changed.

4.3 A Theory of De-commodification of Land Use

The critics of commodification often raise the negative consequences—inequality, for instance—or they call into question what shall be commodified. Sex trade or organ sales (Smith 2009) are typical examples of a usually unwanted form of commodification of the human being. In a famous satirical essay 'A Modest Proposal,' the Anglo-Irish satirist Jonathan Swift (1729) suggested selling the children of the Irish poor as food for the rich. Swift calculated the age in which the selling of children's flesh would be most profitable and presented a list of advantages of such a child-selling policy: 'The poorer tenants will have something valuable of their own, which by law may be made liable to distress and help to pay their landlord's rent, their corn and cattle being already seized, and money a thing unknown' (Swift 1729: 10).

This section presents a theory of de-commodification of land use for people affected by inadequate housing. This does not, however, mean that I regard commodification as something good or bad. The main purpose of the theory is to motivate housing scholars and policymakers to start looking for solutions outside of formal housing markets when they respond to the human right to housing.

My theory draws from Esping-Andersen's 'The Three Worlds of Welfare Capitalism' (2011), first published in 1990. Although Esping-Andersen's three welfare regimes make his publication famous, his book also considers the roots of social policy. Esping-Andersen (2011: 9–16) summarized the explanation models for the welfare state of classical economists such as Alfred Marshall and Karl Polanyi. He explained the positions of liberals as well as Marxist political economists. Regarding Marshall's concept of social citizenship, Esping-Andersen criticized that the author over-emphasized the role of rights:

> The welfare state cannot be understood just in terms of the rights it grants. We must also take into account how state activities are interlocked with the market's and the family's role in social provision. These are the three main principles that need to be fleshed out prior to any theoretical specification of the welfare state. (Esping-Andersen 2011: 21)

Esping-Andersen (2011: 14–16) identified two approaches to explaining the welfare state: the systems or structuralist approach and the institutional approach. The first approach has different explanatory variants. One variant states that social policy became both possible and necessary through industrialization; a second variant sees the rise of modern bureaucracy as the reason, and a third variant regards the welfare state as an inevitable product of the capitalist mode of production (Esping-Andersen 2011: 13–14). The institutional approach emphasizes that the economy should not be isolated from social and political institutions, but should be embedded in social communities (Esping-Andersen 2011: 14–16).

The explanation of the rise of welfare states is only one issue in the welfare states discourse; the other important aspect is the challenge of measuring welfare policies. Esping-Andersen (2011: 18–21) criticized former (quantitative) social policy research attempts that concentrated too much on the level of social expenditure or other economic aspects. He developed a new typology with which he, by regarding OECD countries, differentiated between three forms of welfare regimes: the liberal, the corporatist, and the social democratic regime. His key indicators for differentiating between the different types of welfare states were the rate of de-commodification (of labor), the effect upon stratification, and the qualitatively different arrangements between the state, the market, and the family.

While commodification can be understood as the action of turning something into a commodity, which can be negotiated on the market (see above), de-commodification means the emancipation of individuals from the (labor) market (Esping-Andersen 2011: 22). De-commodification is a process with different roots and 'refers to the degree to which individuals, or families, can uphold a socially acceptable standard of living independently of market participation' (Esping-Andersen 2011: 37). The author had the labor market in mind and discussed the de-commodification of workers, including their needs. He explained the roots of de-commodification as follows:

> In pre-capitalist societies, few workers were properly commodities in the sense that their survival was contingent upon the sale of their labor power. It is as markets become universal and hegemonic that the welfare of individuals comes to depend entirely on the cash nexus. Stripping society on the institutional layers that guaranteed social reproduction outside the labor contract meant that people were commodified. In turn, the introduction of modern social rights implies a loosening of the pure commodity status. De-commodification occurs when service is rendered as a matter

of right, and when a person can maintain a livelihood without reliance on the market. (Esping-Andersen 2011: 21–22)

Stratification, the second aspect, 'refers to the extent to which the welfare state tends to reinforce status conferred by the labor market and can be contrasted to "solidarity" whereby welfare rights are enjoyed equally' (Stephens & Fitzpatrick 2007: 202). Studying social stratification means to identify potential conflict structures (Esping-Andersen 2011: 226). The welfare state is, however, by itself 'a system of stratification [and] an active force in the ordering of social relations' (Esping-Andersen 2011: 23). Stratification refers to inequality. It emphasizes the spatial dimension of housing because 'stratification is reflected in the process of housing allocation' (Hoekstra 2003: 60).

The degree of de-commodification, the effect upon stratification, and the role of the actor groups of family, market, and state (Table 4.3) are the key elements by which Esping-Andersen differentiated between three forms of welfare regimes. In the liberal welfare regime, the market plays a central role by guaranteeing welfare services. The de-commodification rate is low, and the effect upon stratification is high. Social policy only focuses on those 'in current need' (Matznetter & Mundt 2012: 274).

> The consequence is that this type of welfare regime minimizes de-commodification-effects, effectively contains the realm of social rights, and erects an order of stratification that is a blend of a relative equality of poverty among state-welfare recipients, market differentiated welfare among the majorities, and a class-political dualism between the two. (Esping-Andersen 2011: 27)

Table 4.3 Role of actors in the three worlds of welfare capitalism

Role of	Liberal	Corporatist	Social Democratic
State	Marginal	Subsidiary	Central
Market	Central	Marginal	Marginal
Family	Marginal	Central	Marginal

Sources: Esping-Andersen 2011; Gough 2004a: 24

Esping-Andersen (2011: 27) listed the United States, Canada, and Australia as typical examples of this model. In the corporatist regime, rights are attached to class and status; the market plays only a marginal role. The degree of de-commodification is relatively high as well as the effect upon stratification. The key actors along welfare lines are the family and the state. The state's basic task is to subsidize the family, and 'the state will only interfere when the family's capacity to service its members is exhausted' (Esping-Andersen 2011: 27). Austria, France, and Germany were mentioned as typical examples. The Scandinavian countries represent the third type, the social democratic regime. In the social democratic regime, the equality is one 'of the highest standards [and not one] of minimal need' (Esping-Andersen 2011: 27). Consequently, the effect upon stratification is low. The social services are highly de-commodifying and universalistic: 'All benefit; all are dependent; and all will presumably feel obliged to pay' (Esping-Andersen 2011: 28).

The typology was criticized for several reasons on 'a number of fronts' (Dean 2006: 30). Gough (2004a: 24–25) summarized the critics. Most of the critical arguments pointed out the simplicity of the concept. The identification of only three types of regimes only covered 18 countries. Several important aspects have been ignored, for example:

- other components of well-being (besides de-commodification) like autonomy and needs-satisfaction;
- the growing constraints of the global political economy and the growing role of supra-national institutions;
- sources of stratification (other than class analysis) such as religion, ethnicity, and gender;
- the effects of gendered division of labor and household forms; or
- critical social programs like health, education, and housing. (Gough 2004a: 24–25)

The welfare state regime paradigm produced lots of new approaches that added or improved the categorization attempts. Most of the characterizations replied to the simplicity and added a fourth type of welfare state (or more). One modification was the Mediterranean welfare state regime that includes the Mediterranean countries Italy, Greece, Spain, and Portugal (Arts & Gelissen 2002: 142–146). These countries are characterized by a high degree of fragmentation of the social protection systems and relatively little intervention in the welfare sphere (Arts & Gelissen 2002: 145).

The number of publications applying the welfare regime model to housing policies is high, even though scholars mentioned the difficulty of such an analysis (Edgar et al. 2002: 8). The welfare regime models have even been used for explaining the diversity of service provisions, even for homeless people (Hradecký 2008: 178). Housing scholars often used Esping-Andersen's typology to focus on the housing policies of one single European country, like Austria (Matznetter 2002), Sweden (Olsson & Nordfeldt 2008), the Czech Republic (Hradecký 2008), or the Netherlands (Hoekstra 2003 and 2005). Hoekstra (2003: 63) showed that the Dutch housing system includes elements from all three typologies. Some scholars compared different countries with each other, such as Portugal with Wales (Baptista & O'Sullivan 2008), or the Scandinavian countries Denmark, Norway, and Sweden (Benjaminsen & Dyp 2008). Most approaches can be separated in two ways: The researchers either took the categorization of the countries into three or four welfare regime types as a starting point and then evaluated the differences and overlaps (Hoekstra 2005; O'Sullivan & de Decker 2007) or they categorized the countries by using their own housing-related indicators (Hoekstra 2003).

Hoekstra (2005) tested different statistical techniques to determine if there is a correlation between the nature of a society—represented by its welfare state regime—and dwelling types. He differentiated between single-family dwellings and apartments. Hoekstra (2005: 483) found that there are no differences between the percentages of owner-occupied and rental dwellings in the Southern European countries Italy, Spain, Greece, and Portugal, while in the rest of the EU, single-family dwellings are predominately owner-occupied and apartments are rented. Finnish dwellings were regarded as being in the best condition and Portugal was identified as having the most problems (Hoekstra 2005: 484). Hoekstra came to the following conclusion:

> [T]he characteristics and the appreciation of single-family dwellings compared to apartments do not differ much between social-democratic,

> corporatist and liberal welfare state regimes. However, there is a significant difference between these three regime types on the one hand, and the Mediterranean welfare state regime on the other hand. It should be noted that Italy, which is a corporatist welfare state regime according to the Esping-Andersen typology, has much in common with the other Southern European EU countries, which are all Mediterranean welfare state regimes. (Hoekstra 2005: 491)

O'Sullivan and de Decker (2007) analyzed the potentials of the private rental market for homeless people. They tentatively identified four private rented regimes in Europe: Mediterranean countries (Greece, Italy, Portugal, and Spain), in all of which the sector is in decline; liberal regimes (United Kingdom and Ireland), where the sector is growing; social democratic regimes (Denmark, Sweden, Finland, and Norway), which are characterized by a mix of growth and decline; and corporatist regimes (Belgium, France, Germany, and the Netherlands), where, in most cases, a modest decline is evident (O'Sullivan & de Decker: 99). Although rent control for the private rental market was, in its earlier manifestations, seen as an instrument to reduce homelessness, it could also become a barrier for those who search for an accommodation (O'Sullivan & de Decker: 105). Stephens and Fitzpatrick noticed a growing complexity of the housing system:

> The welfare regime will also exert a crucial influence on the scale of homelessness. Welfare regimes that produce high levels of poverty and inequality are likely to have particularly high levels of homelessness because of the relatively weak purchasing power of lower income households in those countries. (Stephens & Fitzpatrick 2007: 208)

Sweden, as a country that is often regarded as a socio-democratic welfare regime, appears to produce relatively low levels of homelessness, while the United Kingdom, as a liberal welfare regime, has more problems with homelessness. Nevertheless, the authors admitted, 'the housing system, both in terms of the operation of the housing market and policy interventions, is capable of either reinforcing or counterbalancing the influence of welfare regimes' (Stephens & Fitzpatrick 2007: 210).

Matznetter and Mundt (2012) summarized the European housing development and differentiated between convergence and divergence approaches. The divergence approach emphasizes the distinctiveness in political ideologies (Stephens & Fitzpatrick 2007: 202) and summarizes three further approaches that focus either on the systematic application of the welfare regime trilogy, on the structure of housing production, or on housing tenures (Matznetter & Mundt 2012: 282). Therefore, the divergence approach allows for placing countries into clusters of regime types (Stephens & Fitzpatrick 2007: 202). The convergence approach is often connected with terms like privatization or transition-to-the-market. Its key message is as follows:

> [W]hile post-war housing needs decreased in quantitative and, later, also in qualitative terms, housing subsidies are being reduced and shifted from the supply side to the demand side of the housing market. Additionally, the social rental housing sector is increasingly superseded by owner-occupation. European integration, the aim to cut public spending, the (in)direct influence of EU competition legislation, and the transition of former communist states to the market fuels the impression of increasing convergence of housing policies across Europe. (Matznetter & Mundt 2012: 282)

Stephens and Fitzpatrick (2007: 201–203) differentiated between the 'old' convergence approach and the 'new' convergence approach. The old convergence approach describes the development of welfare states during the economic growth in the period beginning with industrialization till the oil crisis in 1973. The new convergence approach replies to the consequences of globalization (Stephens & Fitzpatrick 2007: 202). The ongoing trend of commodification in the housing sector has, similar to land titling, been criticized heavily:

> This trend has far-reaching consequences, as the ongoing process of privatisation or conversation of public rental flats into owner-occupied ones is almost irreversible. [. . .] The trend of commodification goes hand in hand with a clear rhetoric that people find upgraded when they own their homes, that they become more proud and willing to work. More

> fundamentally, the trend of commodification is embedded in neoliberal economic theory that the market should allocate as much as possible in the free markets dynamics. (Edgar et al. 2002: 48)

Taking the metaphor of housing as the 'wobbly pillar under the welfare state' (Torgersen 1987) into consideration, one explanation for the 'wobbliness' is this: While the welfare state has its roots in the de-commodification of its inhabitants, particularly the workers, the housing stock is usually under a process of commodification.

Esping-Andersen (2011) looked at old-age pensions, sickness benefits, and unemployment payments when he was measuring de-commodification scores (Matznetter & Mundt 2012: 275–276). The present research places Esping-Andersen's interpretation of de-commodification in the center of attention. As noted, one of my key arguments is that policymakers and scholars who respond to the human right to housing must take the plural meanings of land into consideration. Such a policy works either through the commodification of land or, as I will discuss in the following section, through the **de-commodification of land use.** But first, I give a brief overview of what other scholars understood as de-commodification of housing.

In scholarly discourse, de-commodification of housing means that 'housing outcomes are achieved independently from labour market outcomes' (Stephens & Fitzpatrick 2007: 206) or, as another author worded it, it denotes 'the extent to which households can provide their own housing, independent of the income they acquire on the labour market' (Hoekstra 2003: 60). Hoekstra (2003) examined the housing policy in the Netherlands and transformed Esping-Andersen's two basic indicators of stratification and de-commodification into housing indicators. The housing indicators to measure de-commodification were subsidization and price setting regulation. Hoekstra (2003: 60) acknowledged that, in his model, the responsibility for de-commodification concentrates on the state and ignores the role of other actors. For the state, several opportunities exist, for instance, the regulation of housing prices, production subsidies, or subject subsidization: general income support like pensions, unemployment benefits, or specific subsidies in the field of housing (Hoekstra 2003: 60). Financial measures, however, do not fully emancipate people from markets. Instead, they simply facilitate access. The problem with the aforementioned definitions of de-commodification of housing (Hoekstra 2003; Stephens & Fitzpatrick 2007) is that they focus on the labor market (including its outcomes) and ignore the housing market. De-commodification of housing understood as a process of 'emancipation from the market', must include the housing market. Old-age pensions, sickness benefits, and unemployment payments mean that people can also live adequately without actively participating in the labor market. People are emancipated from labor markets, but there are some difficulties with respect to de-commodification and housing. The high capital intensity of housing (Kemeny 2001: 54) has different causes, and land is probably the most significant. As the availability of land is the first essential condition for a well-functioning housing sector (Angel 2000: 192), it seems that the state, by responding to its obligation to realize adequate housing, has no other choice than to accept the formal land market for housing, even when the state itself functions as an important actor on such a market or tries to facilitate access. The millions of people who are homeless; or live in slums, informal, or illegal settlements are often excluded from such formal housing markets. No one would say that they are emancipated from the markets, and thus, an example of a successful de-commodification of housing. Usually, they are regarded as people affected by inadequate housing because of this exclusion. As millions of people live outside of formal housing markets, I suggest not discussing de-commodification of housing. Instead, we should start thinking about de-commodification of land use for people affected by inadequate housing.

To understand what de-commodification of land use means with respect to the human right to housing, it is of crucial significance to keep in mind that housing is a complex issue that is interrelated with many other rights and needs, such as sleeping, eating, health, and many more. In 2010 and 2011, with a group of first-semester students, I investigated homelessness in Hamburg, Germany (Ajayi et al. 2011). Hamburg is well known for its high land values and its increasing rents. The students interviewed homeless people and examined the causes of homelessness. They came up with interesting findings. Land values and increasing rents were not the main cause for people to become homeless. Usually, people became homeless for several reasons, and each homeless person made his or her own path into homelessness. So, although land values and increasing rents were a high burden to find a flat, many homeless people migrated to Hamburg from other regions of Germany and even from across Europe. The students asked the homeless people why they came to Hamburg instead of choosing other places where it would be much easier to find a roof over their heads. These are two of the most remarkable answers:

- 'In Cologne, I have to pay 50 cents to get a shower; here in Hamburg it's free' (Interview with Torsten, cited from Ajayi et al. 2011: Annex, author's translation).
- 'In Hamburg, everything is for free; that's brilliant' (Interview with Fabio, cited from Ajayi et al. 2011: Annex, author's translation).

Homeless people in Hamburg could not find a flat. Instead, they could satisfy many other needs that are usually connected with housing, such as eating, drinking, washing, doing the laundry, locking up their belongings, or using a computer. Moreover, in Hamburg, homeless people could take of other needs that are not directly connected to housing, such as medical care or legal assistance for free or for a symbolic price. The City of Hamburg has published a brochure where all these facilities are listed (Hamburg 2013). Hamburg was unable to integrate homeless people back into the housing market in a similar way as megacities in the global South cannot offer adequate housing for millions of their inhabitants. However, Hamburg offered several facilities to meet many other needs that are usually met through housing. The Hamburg example shows, from my point of view, what de-commodification of land use for people affected by inadequate housing could look like. I define de-commodification of land use as the extent to which individuals can use the land independently of their income and independently of their involvement in formal housing markets. De-commodification of land use is not the one and only solution for people affected by inadequate housing and shall not stand for an 'everything is for free' policy either. Neither should a policy in the face of de-commodification of land use replace the many housing policies that focus on the reduction of inadequate housing. There are good reasons why scholars propagate 'Housing First' (Busch-Geertsema 2014) approaches for homeless people.

De-commodification of land use, however, stands for the opportunity to have access to land. One example: Wealthy people mostly use the toilet at home (as members of the housing market), at the working place (as members of the labor market), or in a restaurant or other places (as participants in the leisure economy). Homeless people have no home and only seldom participate in formal labor or costly leisure markets. Scholarly discussions on inadequate housing from the perspective of welfare theories often center around the question whether renting or homeownership is the right solution (Section 3.3). In contrast to that, de-commodification of land use for people affected by inadequate housing

- opens the door to regarding inadequate housing not only as a housing problem that can only be solved by housing markets;
- does not buy into the assumption that commodification (in the form of land titling or privatization of the housing stock) is the only solution;
- motivates a focus on land use rights instead of the tenure question;
- gives new inspiration by answering the question of how to deal with informal housing arrangements because it motivates exploration of what land users, such as homeless people or slum dwellers, do; and
- allows people who are outside of formal housing markets to use the toilet and satisfy other needs—without feeling humiliated.

In Chapter 5, I will show that the global attention to the human right to housing has continuously increased. However, the number of people affected by inadequate housing has increased as well—and even faster than the attention did. My theory of de-commodification of land use draws upon my key argument that in the future, we will not be able to completely fulfill the human right to housing. Instead, a land policy in the sense of de-commodification of land use should reduce exclusionary effects for people who could not obtain access to some land uses for financial reasons. A land policy that stresses the principles of de-commodification effects helps the poor to claim some of their citizenship rights to land (B. Davy 2009: 258). Such a land policy emphasizes that the 'everyday social contract of informality' (B. Davy & Pellissery 2013) is as valuable as property rights arrangements in formal housing markets.

I neither want to propose that municipalities and other actors should respond to homelessness only with soup kitchens, rest rooms, or free Internet, nor do I want to suggest mobile toilets as the best solution to improving the living conditions in slums of megacities in the global South. I also do not suggest ignoring the tenure question. However, as proportionally more and more people will live outside of formal housing markets in the future, there are good reasons to start (or keep on) thinking about what de-commodification of land use for people affected by inadequate housing should look like. As Waldron stated, 'Everything that is done has to be done somewhere. No one is free to perform an action unless there is somewhere he is free to perform it. Since we are embodied beings, we always have a location' (Waldron 1991: 296). While German courts deal with the question of whether male tenants' rights include the right to urinate standing up (Section 3.1), de-commodification of land use asks for the places where people affected by inadequate housing can meet needs that are usually met through housing. To not over-theorize the present chapter, I will not fit into the discourse on capabilities or the human development approach (e.g., Nussbaum 2011). However, thinking about the relationship between inadequate housing and capabilities (McNaughton Nicholls 2010) would be a good starting point.

The remaining question is: How can de-commodification in the ICESCR discourse be verified? De-commodification of land use came up in the reports when the states talked about people affected by inadequate housing in the context of many other needs that the states wanted to meet outside of market participation. States that talked about microcredits or land titling (including the involvement of market actors) showed that they entered an economic road to social citizenship. In contrast, when a state described a policy that combined night shelters, soup kitchens, education programs, medical care, or facilities for bathing, washing, and cooking, it showed ingredients of a de-commodification of land use. I expected to find merely traces of de-commodification of land use in the reports. However, to search for such traces was necessary as I wanted to find out if a road to social citizenship exists outside of formal economies—a way of discussion that did not regard an individual affected by inadequate housing as a right holder only for economic reasons.

Part II

The Discourse on the Human Right Housing Under the Monitoring System of the ICESCR

5

Changing Views: Housing in the Past Four Decades

The reporting procedure and the standing of the ICESCR have frequently been the subject of scholarly discourse. So far, however, only a few studies consider the contents of the ICESCR States Parties reports (U. Davy 2013 and 2014) or the contents of the Concluding Observations (Saul et al. 2014; UN-Habitat & OHCHR 2002).

This chapter presents how the view on housing has changed during the 38-year span from 1977 to 2015. The focus is on the different forms of inadequate housing, disadvantaged and vulnerable groups, responding policies, and actors. The chapter describes the narrative of the discourse by showing to what extent ingredients of a globalization of social policy and social citizenship have spread into the states' self-descriptions. I identify traces of de-commodification of land use and describe how the responsibilities have changed. Moreover, I show that discourse events, such as the establishment of the Committee or the publication of the Reporting Guidelines, significantly influenced the reports.

The ICESCR discourse is like a tanker that has continuously been navigating the sea of human rights, but has not changed directions quickly. The reporting of ESC rights started quite leisurely and then continuously improved. However, there is still room for improvement.

M. Kolocek, *The Human Right to Housing in the Face of Land Policy and Social Citizenship*, DOI 10.1007/978-3-319-53489-3_5

The first time period under consideration lasted from 1977 to 1989. This period could be regarded as the experimental phase.

5.1 First Period (1977–1989): The Experimental Phase

Between 1977 and 1989, 60 states (including Czechoslovakia, the German Democratic Republic, the Union of Soviet Socialist Republic, and Yugoslavia) from all UN regions started to submit their first reports. A few states also submitted their second reports. Many states talked about housing in general, but only about half of the 60 states reported inadequate housing. Homelessness appeared in States Parties reports of ten states, though with only little information. Colombia, Iraq, and Tunisia were the only States Parties that talked about street children. When states talked about how to respond to homelessness, they usually mentioned the provision of accommodation. Prevention of homelessness or social support for homeless people played a marginal role if any role at all.

Spaces of Inadequate Housing were more prominent than homelessness: 27 states mentioned this form. Here, too, spatial measures dominated, such as the improvement of infrastructure, slum clearance, or urban renewal. Only five states—Australia, Colombia, Cyprus, Tunisia, and the United Kingdom—reported both forms.

Many states did not focus on inadequate housing. Instead, they talked about housing in a general manner. The story line often ran like this: The country (Austria shall serve as an example) scheduled different housing laws (Austria 1986: para. 48), talked about some financial measures (para. 49–52), and examined protection rights for tenants (para. 55). The country mentioned housing as a demand problem (para. 54), but did not really describe the causes in detail.

More than a fourth of the submitting states concentrated on problems with housing supply which they mainly tried to solve through housing construction programs. Particularly states from the Eastern Block praised their achievements in the building sector: 'In respect of rates of housing construction, the USSR is one of the leading countries in the world' (Union of Soviet Socialist Republics 1980: 16; see also Union of Soviet Socialist Republics 1986: 22). The states praised their achievements by giving statistical information on the high numbers of new flats built: 'Between 1971 and 1977, 922,000 flats were built or modernized, which improved housing conditions for 3 million citizens' (German Democratic Republic 1979: 8). A few states, such as Ecuador (1989: para. 72), Portugal (1983: para. 151), and Poland (1980: 20), recognized urbanization and rural-urban migration as a cause of housing problems.

> Housing is one of the most urgent problems that Ecuador, like all the developing countries, has to tackle. The problem has been aggravated by the mounting pressure on the urban infrastructure on account of the migration from the countryside to the city. (Ecuador 1989: para. 76)

In their reporting, some states seemed to consider housing shortages as unavoidable. Italy (1983: para. 390–398) reported in detail how the crisis of the world economy in the 1970s influenced home-building, and therefore, led to a housing crisis. Moreover, the country explained:

> The housing shortage has also been accentuated by other factors: Internal migration which led to the abandonment of a considerable amount of homes in certain geographical areas, the increase of building in the tourist areas, the use of economic resources for the building of second houses, the curb on mobility within the large metropolitan areas, the transfer of functions in the matter of building to the regions and communes, which in the initial period of activity had to face serious problems of organization and technical preparation, and lastly, the shortage of urbanized areas available for home-building. (Italy 1983: para. 394)

Italy regarded the housing shortage as a problem caused by, inter alia, the scarcity of land ('shortage of urbanized areas for home-building'). The abandonment of homes in certain geographical areas, combined with the increase of building in other (tourist) areas, shows the spatial dimension of the housing shortage. Other states mentioned war or other territorial conflicts as the main causes of inadequate housing. Cyprus, for instance, reported homelessness and other forms of inadequate housing as caused by the 'Turkish invasion' in 1974:

> As a result of these events the right to housing has suffered a serious regress, because Turkey not only alienated 200,000 people from their legitimate houses, but also put a halt to an otherwise strong private housing market. Investments in housing, in the period after the invasion, under unfavourable expectations, uncertainty and seriously limited financial and real resources dropped to the lowest ever recorded point. (Cyprus 1979: 18[1])

Eviction and expropriation also played a marginal role. Nine countries mentioned eviction and fifteen countries mentioned expropriation in their reports (Canada, India, and Denmark did both). Most states that did so talked about prevention and compensation measures. Privatization in the housing sector did not come up, neither did informality. Although a small number of states talked about supporting home-ownership, tenants' rights and construction of and financial support for rental accommodation was mentioned more often.

Financial measures, such as housing loans, were mentioned for groups with special needs, such as families (Czechoslovakia 1980: 5) or students (Yugoslavia 1982: 23). Usually, the states regarded housing as a state's task, with some exceptions: 'The State should play a secondary role in housing. It is for the private sector to marshal resources and means to meet aspirations for housing' (Chile 1979: 6). 'In Cyprus, with its long tradition of private housing market, there are no specific laws and regulations designed to promote the right to housing' (Cyprus 1979: 19).

The variety of the mentioned actors in the sections concerning inadequate housing was low. On average, the states reported the involvement of 1.7 groups of actors (of seven) regarding the human right to housing. While national governmental actors dominated,

[1] In reports that are not structured by paragraphs (which applies mostly to the first-generation reports), I have quoted the page number.

NGO+ and supranational actors were hardly mentioned by the states. In detail, that means 24 of the 32 (75%) states that reported inadequate housing gave an account of an involvement of governmental actors. The likelihood of mentioning any of the other groups of actors was less than 25%. Two states mentioned an involvement of NGO+, while five states mentioned supranational actors. This does not mean, however, that supranational actors did not come up in the reports at all. They were just mentioned less often with respect to inadequate housing than in the sections concerning other rights. Seven states mentioned the involvement of the family or self-help as an actor, and six states talked about local actors. The states reported self-help or the family when they reported on housing in rural areas (e.g., Portugal 1983: para. 162; United Republic of Tanzania 1979: para. 7).

For the most part, the states brought up inadequate housing in a few sentences or paragraphs. A few states—namely Australia, Cameroon, Canada, Cyprus, Ecuador, India, the Netherlands, Portugal, and Tunisia—reported on inadequate housing in a broader scope. These reports had different focal points. Problem groups and housing problems in rural areas were largely the centers of attention. Australia (1980), for example, talked about housing policies for refugees, Aboriginals, veterans, emergency housing for women and youth, or housing for aged and disabled people. Housing, it seemed, was regarded as a problem, especially for people with special needs.

Canada gave detailed information on measures taken or envisaged to solve the special problems of housing, water supply, and sanitary conditions in its rural areas 'where some of the poorest housing and living conditions remain' (Canada 1983: para. 64). India (1983: para. 139–146) also explained in detail how the country improved the access to water and sanitation, especially for its rural population. Portugal (1983: para. 161) explicitly described rural housing as an important problem. While the term urban was frequently linked with the terms development or renewal, the term rural was more likely connected with the terms problem, poor, and poverty. However, there were exceptions. Western European countries such as the Federal Republic of Germany (1979: 31), Denmark (1980: 17), and the United Kingdom (1980: 35) explicitly stated that they had no specific problems with housing in their rural areas.

In 1986, the **Committee** on Economic, Social and Cultural Rights was established (2.3). Till then, as Craven (2002: 1) claimed, the Covenant 'existed only as textual reference point subject to the speculative claims of both its proponents and detractors'. Since 1987, the Committee started to respond to the States Parties reports in the form of Concluding Observations or Comments.[2] Until 1989, the Committee held three annual sessions and published the results each year in a report. The annual reports contained, among other things, organizational and other matters, the first General Comment, provisional Rules of Procedure, statistical information concerning the status of submission of the countries, and the Committee's consideration of some States Parties reports, albeit not of all states (CESCR 1989). The annual reports' main goal seemed to be to clarify the Committee's requests for the contents of the States Parties' reports and to develop a deeper understanding of the relevant issues with respect to specific rights, such as the right to an adequate standard of living (CESCR 1989: para. 310). In these Concluding Observations, housing in general, as well as inadequate housing, played only a marginal role.

All in all, the first reporting period could be called an experimental phase. This applies to the States Parties reports as well as to the Committee's Concluding Observations. Affordability and access to basic amenities (in the rural sectors) dominated. The main housing topic was 'supply.' When considering the states' talk through the 'respect-protect-fulfill' lens (Maastricht Guidelines 1997), it could be said that the fulfill aspect dominated the view on housing. The reporting states treated housing as a spatial issue that should be met through governmental mass construction programs and financial investments.

[2] Until 1992, the Committee's comments to the States Parties reports were included in the Committee's annual reports. Since 1993, the Concluding Observations have followed the same principles as those of other human rights committees. As Craven noted, '[these] new concluding observations, while by no means perfect, represent an important improvement both in terms of the level of detail provided and in the quality of assessment' (Craven 2002: 88). In the following, my use of the term Concluding Observations will also include the Committee's comments to the States Parties reports from 1987 to 1992.

If housing was called a problem, it was either a problem of special groups in need, a rural problem, or both. The states showed ingredients of social citizenship only as they indirectly touched upon issues of inequality in the context of housing, for instance, by reporting on such marginalized groups. While Spaces of Inadequate Housing were mentioned in the reports of nearly half of the reporting states, homelessness played a secondary role. Housing was a national issue. As housing was often reported as a rural problem, the group family/self-help was the second most frequently mentioned group of actors. Governmental actors dominated in the states' reporting, and supranational actors, such as the World Bank, were underrepresented. Global influences were mentioned in the paragraphs concerning causes, for instance, because states reported territorial conflicts that caused inadequate housing, or when they reported housing in the context of a global financial crisis. Although financial measures were among the most prominent policies, market actors did not play an important role. Many aspects of housing that nowadays seem to be evident in actual housing discourses like informality, eviction, land titling, or privatization played, if at all, a marginal role.

5.2 Second Period (1990–1999): Many Changes, Many Challenges

The second period is influenced by the consequences of the dissolution of the Soviet Union and the Socialist Federal Republic of Yugoslavia. The former Soviet republics, as well as the former members of the Socialist Federal Republic of Yugoslavia, ratified the ICESCR quickly after gaining their independence. In sum, 50 states ratified the ICESCR between 1990 and 1999, and 74 states (including Yugoslavia) submitted at least one report. The Committee published Concluding Observations to States Parties reports of 72 States Parties. In the beginning of the 1990s, the member states submitted both types of reports: reports submitted in three stages and full reports. In 1993, the situation (finally) became more straightforward. All submitted reports were full reports, and the Committee started to respond to each report with Concluding Observations published in a separate document.

The attention to housing increased. The Committee's publications and the states' reporting became more multifaceted by considering several aspects. Less than 20% of the 74 submitting states neither mentioned homelessness nor Spaces of Inadequate Housing. A total of 70% reported homelessness, nearly 60% reported Spaces of Inadequate Housing, and almost half (34 states) of the 74 reporting countries reported both main forms; 34 states (46%) mentioned problems with their housing supply, so the attention to housing shortages was also still increasing. In contrast to the first period, not only the attention to inadequate housing increased, but the mode of talking about both forms changed fundamentally. People affected by inadequate housing were no longer regarded as an anonymous mass.

Homelessness began to be increasingly a social problem. Accordingly, the states talked about the combination of different measures. Construction programs and financial measures were no longer the only solutions mentioned. Prevention of homelessness and social support appeared as new policy goals, for instance, in Canada's report:

> It is recognized that the needs of the homeless are very closely associated with other social and health needs. The homeless require more than just shelter assistance. They required (sic) permanent accommodation and social services to break the cycle of unemployment and subsequent social problems. (Canada 1991: para. 113)

Street children were also more likely to be mentioned (32%) than in the first period. However, they were often not discussed in the sections concerning Article 11 ICESCR. Instead, the states considered them as a family rights issue (Article 10 ICESCR) (e.g., Iceland 1997: para. 100; Mexico 1997: para. 232–236; Sri Lanka 1996: para. 201). Homeless people were often regarded as one problem group of many other vulnerable and disadvantaged groups (e.g., Algeria 1994: para. 145; Finland 1995: para. 286–293; United Kingdom 1996: para. 187). Sometimes, the group of homeless people was divided into other sub-groups: Street children, homeless students, homeless juveniles, or homeless mothers.

Even homeless teachers (Republic of Korea 1993: para. 513) were mentioned. Many states no longer understood homelessness in a simple manner. Instead, they tried to capture its complexity. Denmark used the term socially excluded to summarize different vulnerable and disadvantaged groups:

> Although Denmark can boast a fine-meshed social and health service safety net, there are people to whom society has been unable to offer adequate help. Included in this small group are homeless people, drug and substance abusers, street children, a few mentally ill patients, prostitutes and some immigrants and refugees. [...] The homeless category has changed in recent years: there are more young people, women, drug and substance abusers, people on anticipatory pension, mentally ill patients and refugees. There are provisions for the socially excluded in the form of residential homes, institutions and shared housing. Improvement in their condition is a top political priority. (Denmark 1996: para. 143)

The quotation shows that once a country started to regard homelessness as more than a mere housing problem, the country's language showed a kind of helplessness. Denmark seemed to ascribe various responsibilities for the problem: Society had not been able to offer adequate help. The country seemed to be confronted with the newly recognized complexity of the category of homelessness, which made it difficult to respond to the problem sufficiently. The solution mentioned in the report was to provide the 'socially excluded' with residential homes, institutions and shared housing, and to improve their condition. Denmark regarded itself as a country with a 'fine-meshed social health service net' and called the group a 'small' one, without confirming its size with any exact statistical information. The recognition of some groups affected by inadequate housing as 'socially excluded' indicates, however, that exclusion was recognized as a problem. Some states started to define homelessness or talked about their concept of homelessness:

> In Finland, persons who have turned to the authorities because they are without accommodation or persons known by the authorities to be without accommodation, as well as persons residing in institutions or like

> accommodation because of the lack of suitable housing or non-institutional services ('hidden homelessness') have been included in the concept of homelessness. (Finland 1995: para. 305)

Finland's inclusion of 'hidden homelessness' into the country's 'concept of homelessness' once again demonstrates the newly recognized complexity. The language also changed. The States Parties started, albeit seldom and carefully, to talk about rights of the affected persons, duties, and obligations for the governments or local authorities (e.g., the Netherlands 1996: para. 134; United Kingdom 1996: para. 163).

In the 1990s, even though the attention to Spaces of Inadequate Housing decreased in relation to the attention to homelessness, this form was also discussed in a more complex manner. First, the range of the mentioned forms of (and terms for) Spaces of Inadequate Housing rose. New forms came onto the agenda, such as illegal settlements, illegal housing sectors, campsites, cage homes, or makeshift dwellings. Reports of squatting, squatter settlements as well as the term illegal became more prominent. The term illegal settlements was explicitly asked for in the revised General Guidelines by the Committee (CESCR 1991b: 13), and this could explain why, in the context of housing, illegal was more likely mentioned than the term informal. In a word count supported with the software ATLAS.ti, I compared the ratio between the lexemes illegal and informal.[3] The ratio was 91% (illegal) to 9% (informal).

In the 1990s, the states' attention to the urban poor rose, and urban poverty often was automatically linked with inadequate housing:

> The urban poor live in blighted areas with substandard housing and insufficient urban infrastructure, such as roads, city water and sewerage. However, it was difficult for the residents themselves to improve the housing conditions. They lacked financial resources due to their low incomes, and the physical conditions of their houses and settlements did not meet the standards required by official urban plans and building

[3] N = All terms *illegal, informal, illegality, informality* mentioned in the housing relevant paragraphs of the States Parties reports from 1990 to 1999.

> codes. These problems aggravated the poor quality of the living environments. (Republic of Korea 1999: 242)

The quotation illustrates that self-help was regarded as an insufficient way to reach adequate housing. The trend to regard housing problems as urban problems does not indicate that rural housing problems no longer played a role. As for many other trends, too, there were exceptions. One such exception is Sri Lanka (1996: para. 226), a state that still regarded housing poverty predominately as a rural phenomenon.

The range of the reported measures increased, too. In contrast to the language used to discuss homelessness, the language used to discuss policies for Spaces of Inadequate Housing did not become more socially oriented. Instead, the states started to combine spatial solutions with financial measures or land titling. Although squatters had usually been regarded as rights violators, a few states began to recognize the illegal occupation of land and reported land titling. Argentina (1997: para. 193), for instance, listed legislative measures to grant land titles to segments of the population living in illegal settlements; similarly, Colombia (1994: para. 548) and Mexico (1992: para. 201). Israel (1998: para. 474 and 479–482) reported its policies to legalize Bedouin settlements. The Philippines mentioned an involvement of the World Bank in the state's titling policy:

> [A] World Bank shelter sector loan for the same period was utilized to rehabilitate the country's housing mortgage institution. A portion of it funded the Community Mortgage Programme which was launched in 1988 as an innovative programme to enable squatters and slum dwellers to own the lands they occupy. (The Philippines 1994: para. 391)

The number of mentioned actors increased, too. The states mentioned the involvement of 1.7 groups of actors on average in their paragraphs concerning inadequate housing in the first period, whereas in the second period the rate was 2.6. In more detail: Of the 74 states that submitted at least one report 61 (82%) talked about at least one main code of inadequate housing, 34 states (46%) did both; 49 of these 61 states (80%) talked about an involvement of governmental actors, 25 states (41%) mentioned local actors, 24 states (39%) mentioned NGO+, 19 states (31%) mentioned market actors and 11 states (18%) mentioned supranational actors (other actors: 20 states, 33%). Eight states (13%) were reporting self-help or family involvement as ways to reach adequate housing. The numbers show that governmental actors were still dominating, but usually, one or two actors from the other six groups were also mentioned when the states reported their policies designed to accomplish adequate housing.

Of 74 states, 34 (46%) addressed refugees, but only half of them linked the group explicitly with housing. One example was Denmark:

> The most comprehensive integration effort is directed towards refugees. Recognized refugees and, in certain circumstances, their close relatives are offered an integration programme organized on behalf of the State by the Danish Relief Council, a private humanitarian agency. The Danish Relief Council provides housing for refugees in all parts of Denmark, in addition to social counselling and assistance. (Denmark 1996: para. 148)

Denmark showed here, by talking about its 'comprehensive integration effort', ingredients of social citizenship. The other three disadvantaged and vulnerable groups were less likely to be mentioned. 28 States Parties (38%) talked about indigenous people, 12 of them (43%) linked this group with housing or land issues. This means that 57% of the states that spoke of indigenous people did not explicitly link this group with housing or land questions, but with other rights, for instance, with the right to education (Sri Lanka 1996: para. 346) or the right to health (Venezuela 1998: para. 227).

> The highest percentage of unemployment was found among the early school leavers who were either unqualified or unskilled to fill vacancies. Apart from this group, the indigenous peoples who live in the hinterland areas are considered to be more disadvantaged because of their location. However, the increased economic activity, especially mining and logging in these areas, has resulted in increased jobs for residents. (Guyana 1995: para. 17)

In this quotation, the country regarded the location as one main problem that at least could be solved through 'increased economic activity'. The economic potentials of land—namely, its mineral resources—should reduce unemployment. The country regarded land as a double-edged sword. On the one hand, the location 'in the hinterland areas' is a reason why indigenous people were regarded as disadvantaged; on the other hand, its economic potentials as a mineral resource should, so reported Guyana, solve the location problem.

Roma—who were hardly mentioned in the first period—appeared on the agenda of seven states, but were only connected with housing by four states (e.g., Portugal 1997: para. 42–47; Spain 1994: para. 101).

The **Committee's role** in the discourse changed considerably, too. Since the beginning of the 1990s, the Committee became an active and influential participant. In this second period, housing was the most prominent right of the ESC rights in the Committee's publications. The Committee published two General Comments concerning housing (CESCR 1991a and 1997). The first General Comment's purpose was to clarify the dimension of adequate housing and the second General Comment observed and reacted to a growing violation of the States Parties' obligations, that is, forced evictions.

In its General Comment No. 4 (CESCR 1991a), the Committee concretized its consideration of the right to housing. The Committee suggested that the states should be required to adopt a national housing strategy (CESCR 1991a: para. 12) and interpreted the formulations of Article 11, para. 1, ICESCR anew:

> While the reference to 'himself and his family' reflects assumptions as to gender roles and economic activity patterns commonly accepted in 1966 when the Covenant was adopted, the phrase cannot be read today as implying any limitations upon the applicability of the right to individuals or to female-headed households or other such groups. Thus, the concept of 'family' must be understood in a wide sense. (CESCR 1991a: para. 6)

The Committee suggested viewing housing as 'the right to live somewhere in security, peace and dignity' (CESCR 1991a: para. 7) and elaborated upon the idea that housing adequacy depends on social, economic, cultural, climatic, ecological, and other factors (CESCR 1991a: para. 8). The Committee identified seven basic aspects to be considered. Table 5.1 illustrates these aspects, including a short summary of the Committee's explanations. The Committee's list has been repeated, substantiated, shortened, modified, and commentated on in other sources as well (e.g., Craven 2002: 335–347; UN-Habitat 2009a: 4).

General Comment No. 4 opened the door to regarding housing in a more multifaceted manner. Table 5.1 illustrates that housing is related to many other needs and rights and shall not be viewed in isolation. The seven aspects are sometimes closely linked with each other. Availability of services (access to natural and common resources), for instance, is always closely linked to location aspects and the location is linked to housing costs (affordability). Some of the aspects of Table 5.1 are easy to

Table 5.1 Aspects of adequate housing

Legal security of tenure	Tenure takes a variety of forms. All persons should be protected against forced evictions, harassments and other threats.
Availability of services, materials, facilities, and infrastructure	An adequate house must contain facilities essential for health, security, comfort, and nutrition. Access to natural and common resources, safe drinking water, energy, heating and lighting, sanitation and washing facilities, means of food storage, refuse disposal, site drainage, and emergency services.
Affordability	The percentage of housing cost should be commensurate with income levels.
Habitability	Inhabitants must be protected from bad weather effects and other threats to health.
Accessibility	Disadvantaged groups must be accorded full and sustainable access to adequate housing, and their special needs should be considered.
Location	Access to employment options, health care services, schools, child care centers, and other facilities. Housing should not be built on polluted sites.
Cultural adequacy	Modernization should not sacrifice the cultural dimensions of housing.

Source: CESCR 1991a: para. 8.

quantify (e.g., affordability), while others are more difficult to value objectively (e.g., location). The reader may ask how many aspects must be violated to identify a housing situation as inadequate. UN-Habitat (2009a: 4) emphasized that the housing situation is inadequate when at least one of the aspects is violated.

The middle of the 1990s was characterized by some frustration regarding the many reports that were delayed or not submitted, and violations of ESC rights. General Comment No. 7 (CESCR 1997) was the first General Comment issued by the Committee to address a violation of the rights in the Covenant. The motivation for the General Comment was that the Committee had considered a significant number of forced evictions in the recent years (CESCR 1997: para. 1). The Committee defined forced evictions as 'the permanent or temporary removal against their will of individuals, families and/or communities from the homes and/or land which they occupy, without the provision of, and access to, appropriate forms of legal or other protection' (CESCR 1997: para. 4). It further noted a reduced involvement of the state:

> [I]n view of the increasing trend in some States towards the Government greatly reducing its responsibilities in the housing sector, States parties must ensure that legislative and other measures are adequate to prevent and, if appropriate, punish forced evictions carried out, without appropriate safeguards, by private persons or bodies. (CESCR 1997: para. 9)

Although stressing that sometimes evictions may be justifiable (CESCR 1997: para. 12), eviction should not result in rendering individuals homeless (CESCR 1997: para. 17). While General Comment No. 4 is one of the most cited sources in discourses on the content of the human right to housing, Leckie stated that no measure has been as significant as the adoption of General Comment No. 7: 'This general comment provides the most authoritative pronouncement on forced evictions under international law' (Leckie 2001: 165).

Between 1990 and 1999, the Committee published more than 100 Concluding Observations responding to the States Parties reports of 72 states. This includes also the Concluding Observations for states that did not submit any report (see also Table 2.2, p. 36). With respect to Guinea, a state that has yet to turn in any report, the Committee noted:

> The purpose of the reporting system established by the Covenant is for the States Parties to report to the competent monitoring body, the Committee on Economic, Social and Cultural Rights, and through it to the Economic and Social Council, on the measures which have been adopted, the progress made, and the difficulties encountered in achieving the observance of the rights recognized in the Covenant. Non-performance by a State party of its reporting obligations, in addition to constituting a breach of the Covenant, creates a severe obstacle to the fulfilment of the Committee's functions. Nevertheless, the Committee has to perform its supervisory role in such cases and must do so on the basis of all reliable information available to it. (CESCR_GIN 1996: para. 3)

The Committee is free to raise issues where it deems it necessary with respect to a State Party report. The number of housing topics in the Concluding Observations was significantly lower than in the States Parties reports. Of the 72 states, 36% were questioned by the Committee with respect to homelessness; in the case of Spaces of Inadequate Housing, it was 31%. For more than half of the addressed states, neither homelessness nor Spaces of Inadequate Housing were touched upon. Street children were mentioned in Concluding Observations addressed to 14 states (19%). The attention (understood as the number of States Parties that were tackled by the Committee) given to other housing topics (for instance refugees, asylum seekers, eviction, housing shortage, and so on) was also much lower than in the States Parties reports.

The Committee's responding strategy could be described as follows: The Committee reacted to each state individually, depending on the political and socio-ecological circumstances in the country. I show three examples. In the case of Afghanistan, the Committee took a special interest in the problem of five million Afghan refugees (CESCR_AFG 1992: para. 92). By responding to a report from Argentina, the Committee commented on the state's education policy for indigenous people after the economic crisis and asked, among other things, whether any international cooperation and solidarity had occurred since Argentina's restoration of democracy (CESCR_ARG 1990: para. 236–237). Third, the Committee paid almost as much attention to eviction as to homelessness and Spaces of Inadequate Housing. This is remarkably different from the states' degree of attention paid to evictions and proves once more the Committee's frustration concerning the states' eviction policies. Sometimes, the Committee reminded the countries to pay attention to its General Comments (e.g., CESCR_MEX 1994: para. 13) and bolstered the view expressed there as follows:

> The Committee is very concerned about the large number of illegal occupations of buildings, particularly in Buenos Aires, and the conditions in which expulsions are carried out. The Committee draws the attention of the Government to the full text of its General Comment No. 4 (1991) on the right to adequate housing and urges the Government to ensure that policy, legislation and practice take due account of that General Comment. (CESCR_ARG 1994: para. 237)

In the presented quotation, the Committee regarded itself not as concerned about expulsions per se, but the conditions in which expulsions were carried out. The next quotation shows in more detail what a policy for adequate housing could look like:

> The Committee recommends that Belgian authorities take appropriate measures to promote investment programmes and encourage, in particular, the construction of low-cost rental housing. In this connection, the Committee refers to its General Comment No. 4 of 1991 concerning the right to adequate housing and points out that in situations indicating a clear deterioration in the enjoyment of that right, urgent measures should be taken by the authorities concerned. The Committee, in view of the problems in the housing sector, which are still considerable, urges the Government to establish an official, nationwide Commission on Housing, comprised of representatives of Government, non-governmental organizations and other relevant groups. In view of the non-discrimination clauses contained in article 2 (2) of the Covenant, the Committee strongly urges the Government to fully ensure that persons belonging to ethnic minorities, refugees and asylum seekers are fully protected from any acts or laws

> which in any way result in discriminatory treatment within the housing sector. In view of information received by the Committee that not all social housing units are occupied by lower income groups, the Committee urges the Government to undertake all necessary measures to ensure that lower-income groups have access to social housing which is affordable. (CESCR_BEL 1994: para. 14)

The Committee here reminded the state to observe General Comment No. 4 and deemed it necessary to urge the state to establish a 'Commission on Housing' that included representatives of the government, NGOs, and other relevant groups. The Committee explicitly mentioned NGOs, but no market actor. Instead, the state should decide what 'other relevant groups' should participate in the commission. Moreover, the Committee stressed the obligation based on Article 2, para. 2 of the Covenant and 'strongly urge[d]' the government to include ethnic minorities, refugees, and asylum seekers in its consideration. Additionally, the Committee criticized that not all social housing units were occupied by low-income groups. The Committee seemed to encourage the state to explicitly adopt more equality in its housing policies.

In this second time period, the Committee was more self-confident and did not shy away from housing problems that, from its perspective, were caused by territorial conflicts. The Concluding Observations for an Israeli State Party report prove that well:

> The Committee deplores the continuing practices of the Government of Israel of home demolitions, land confiscations and restrictions on family reunification and residency rights, and its adoption of policies which result in substandard housing and living conditions, including extreme overcrowding and lack of services, of Palestinians in East Jerusalem, in particular in the old city.
>
> The Committee notes with concern the situation of Arab neighbourhoods in mixed cities such as Jaffa and Lod which have deteriorated into virtual slums because of Israel's excessively restrictive system of granting government permits without which it is illegal to undertake any kind of structural repair or renovation.

> The Committee notes that despite State party's obligation under article 11 of the Covenant, the Government of Israel continues to expropriate Palestinian lands and resources for the expansion of Israeli settlements. Thousands of dunams (hectares) of land in the West Bank have recently been confiscated to build 20 new bypass roads which cut West Bank towns off from outlying villages and farmlands. The consequence—if not the motivation—is the fragmentation and isolation of the Palestinian communities and facilitation of the expansion of illegal settlements. The Committee also notes with concern that while the Government annually diverts millions of cubic metres of water from the West Bank's Eastern Aquifer Basin, the annual per capita consumption allocation for Palestinians is only 125 cubic metres while settlers are allocated 1,000 cubic metres per capita. (CESCR_ISR 1998: para. 22–24)

Here again, the Committee's main concern was the unequal treatment of different groups affected by inadequate housing. The Committee also used and quoted from other sources to address grievances. It claimed to have found a lack of information in the States Parties reports, or when a state had not submitted any report, the Committee used alternative sources of information, such as the United Nations Human Development Program (UNDP) to support its concerns (e.g., CESCR_GMB 1994: para. 6; CESCR_ LKA 1998: para. 15).

The Committee even decided to send two of its members to Panama to pursue the dialogue directly with the Government with a clear focus on the right to adequate housing (CESCR_PAN 1994: para. 356–358). In preparing the mission, the Committee received further information from several United Nations organs (e.g., UNDP and HABITAT), specialized agencies (ILO and IMF), regional organizations, and non-governmental organizations (CESCR_PAN 1995: para. 13). In paragraphs 17–70 of the Concluding Observations (CESCR_PAN 1995), the findings of the mission were reported. The Committee demonstrated a skeptical perception of the role of private actors with regard to housing policy:

> The private sector, which is represented mainly by the Caja Panameña de la Construcción (Panamanian Construction Company) (CAPAC) is

> involved in housing and road construction, but mainly builds high-cost and medium-range individual and co-owned dwellings and dwellings in the lower-medium price range. It is quite clearly subject to the play of market forces. It depends on the financial policy of the commercial and mortgage bank and aims to meet only 'genuine' demand, that is to say, from persons who need a dwelling and can pay for it. It considers that it is the Government's responsibility to build housing for the poorest population groups. (CESCR_PAN 1995: para. 30)

The Committee did not prohibit an involvement of market actors. However, when it came to inadequate housing or housing 'for the poorest population groups', the Committee strongly reminded the state that it regarded the government as the responsible actor. In summary, the Committee did not bring up the subject of privatization in the Concluding Observations often, but when it did, the Committee was skeptical:

> The Committee, while acknowledging the State party's need to raise financial resources to subsidize its economic reform and development programme, expresses its concern about the Government's plans to privatize communal land with a view to making it accessible for commercial use and urban development. The Committee recalls that approximately 90 per cent of the land in Solomon Islands is held under customary land tenure, meaning that the land belongs to the community as such rather than to individuals. The Committee would like to draw the Government's attention to the fact that the envisaged privatization of land under customary tenure may undermine the foundations of Solomon Islands' society and could lead to the dispossession of the majority of people, thereby depriving them of their basic source of income. With regard to the Government's plans to privatize housing completely, the Committee is concerned that the number of homeless people in the urban areas will increase considerably. (CESCR_SLB 1999: para. 20)

The Committee's view on commodification in housing policies seems to be somehow contradictory. On the one hand, the Committee praised the states by fostering their titling plans and programs (e.g., CESCR_ARG 1994: para. 228; CESCR_DOM 1994: para. 5); on the other hand, the Committee was often skeptical when market forces were too dominant (e.g., CESCR_NIC 1994: para. 6; CESCR_PAN 1995: para. 30; CESCR_SLB 1999: para. 20). However, perhaps this contradiction was the result of the strategy to respond to the States Parties reports and the specific situation in each country individually. Moreover, it seems that the Committee accepted the involvement of market actors if the poorest population groups were not affected negatively. Equality goals, or at least the decline of inequality, were central purposes in the Committee's Concluding Observations.

All in all, the second period differed considerably from the first one for several reasons. The states perceived inadequate housing increasingly as an urban problem linked with many other rights and needs (e.g., family rights, the right to health, or the right to social security). In relation to Spaces of Inadequate Housing, homelessness received a little bit more attention. The states recognized the complexity of the right to housing by differentiating between several problem groups or types of Spaces of Inadequate Housing, as well as by bringing up new measures to eradicate homelessness and Spaces of Inadequate Housing. While the reporting considering Spaces of Inadequate Housing concentrated on economic aspects, the perception of homelessness seemed to move in a more social direction. Homelessness was no longer discussed merely as a housing problem, but as a problem related to family rights (when talking about street children), the right to work, or the right to health. The way in which the states started to talk about homelessness showed ingredients of social citizenship because the states emphasized re-integration goals. In contrast, Spaces of Inadequate Housing were increasingly regarded as an urban problem that still required spatial measures, sometimes combined with land titling.

The range of the actors mentioned in the paragraphs concerning inadequate housing was significantly higher than in the first period. The right to housing was no longer regarded as a task only for governmental actors on the national level, but also for NGO+, market actors, and sometimes also supranational actors. In addition, as the problem of inadequate housing had become more urban, more local actors were involved, while the rate of mentioning family and self-help declined. One could say that in the language of the Maastricht Guidelines (1997), inadequate housing was no longer regarded mainly through the state's fulfill lenses. Social protection played an increasingly more important role.

The Committee significantly influenced the changing view with two General Comments related to housing and its continuous emphasis on housing rights in the Concluding Observations. By doing so, the Committee showed a skeptical perception of market forces in the field of housing and critically questioned states about their unequal treatment of different groups affected by inadequate housing. The Committee strongly reminded the states that governmental actors should be chiefly responsible for the poorest groups. The second period demonstrated the difference between the states' self-description and the Committee's positions. In this second period, the states began to recognize people affected by inadequate housing as citizens. In the language of the Maastricht Guidelines (1997), the protection aspect seemed to dominate.

5.3 Third Period (2000–2015): The Rise of Global Social Citizenship

Between 2000 and 2015, 21 new states ratified the ICESCR and 118 states submitted at least one report. In this period, once again, the attention to inadequate housing increased. Of the 118 reporting states, 79% brought up homelessness, 47% mentioned street children, and 77% talked about Spaces of Inadequate Housing. Once more, the degree of attention given to homelessness was somewhat higher. However, 64% reported both main forms, and only 8% of the reporting states mentioned neither homelessness nor Spaces of Inadequate Housing, meaning that 108 states talked about at least one of the two forms. Street children or homeless juveniles were again more likely regarded as a family rights (Article 10 ICESCR) concern. The awareness of housing shortages was again 46%. Although the discussion in the 2000s was different than in the 1990s, the differences were not nearly as immense as they were between the first and second period. In general, trends that emerged in the second period continued or were even strengthened from 2000 on. I assume that many states (particularly the older ICESCR member states) developed a sort of reporting routine with the Committee. The discipline regarding reporting grew, even though there is still room for improvement. Moreover, the discourse gained more elements of a contentious debate between the States Parties and the Committee.

Although the states' attention to housing issues continuously increased, there are some states that spent less space in the report on the topic in the last report than in the previous reports. France, for instance, reported inadequate housing in detail in its second (France 2000) and third (France 2007) periodic reports, but spent considerably less space on it in its fourth report (France 2013). Iceland (2010: para. 202) acknowledged that, since the country's last periodic report, no fundamental changes had taken place in its legislation and regulations concerning housing.

A lot of states, often from Africa, explained in detail how they tried to deal with poverty (e.g., Angola 2014: para. 171–187; Burundi 2013: para. 219–220; Sudan 2012: para). Many states discussed poverty in its full complexity, meaning that they looked upon many rights such as the right to food, housing, health care, or education. By reporting the causes of poverty, Sudan mentioned the 'impact of economic liberalization on poor and vulnerable population groups, the economic sanctions impeding access to international initiatives, the protracted civil wars and the mounting external debt' (Sudan 2012: para. 78), and Togo (2010: para. 532) the damaging effects of structural adjustment, and the lack of a national housing policy. Some states mentioned rural poverty (e.g., Armenia 2011: para. 256; China 2010: p. 82) or poor rural women (e.g., Egypt 2010: para. 381), but only a few talked about urban migration as a cause for housing problems (e.g., Kenya 2013: para. 157).

Armenia reported that the country reviewed their 'Poverty Reduction Strategic Programme' in 2008 and renamed it as the 'Sustainable Development Programme' (Armenia 2011: para. 250). Sustainable development as a term as well as policy goals under the umbrella of sustainability became more prominent, including those related to housing (e.g., Indonesia 2012: para. 201; Vietnam 2011: para. 335–336).

China presented a remarkable policy that combined a housing policy that was prominent in the first phase (mass construction by the state) with sustainability goals—which are typical for the new Millennium:

> The Chinese Government has accelerated the construction of housing, centring on the two key themes of 'adequate housing for all' and 'sustainable development of human settlements in an urbanizing world'. (China 2010: p.51)

One of the key questions in discourses on sustainability is how to deal with water. In 2002, the Committee published a General Comment on the right to water (CESCR 2002a). In this General Comment, the Committee explicitly stated that

> [d]eprived urban areas, including informal human settlements, and homeless persons, should have access to properly maintained water facilities. No household should be denied the right to water on the grounds of their housing or land status. (CESCR 2002a: para. 16c)

The Committee differentiated between two main forms of inadequate housing, namely informal urban settlements as an example for a deprived urban area on the one hand, and homelessness on the other hand. The General Comment significantly influenced the states' reporting. Since 2002, they reported on their policies for the right to water often in great detail (e.g., Belarus 2010: para. 339–35; Belgium 2010: para. 247–248).

By considering inadequate housing, the boundaries between homelessness, Spaces of Inadequate Housing, and other housing subjects blurred. States that reported homelessness and Spaces of Inadequate Housing did so simultaneously and also reported on other topics in the same paragraphs, such as eviction or expropriation. Morocco (2013: para. 175) distinguished between homeless people and people living in unhygienic housing, who were then classified in shantytowns, substandard housing, and dangerously run-down housing. In sum, the subject density increased:

> In general, people who are evicted from their homes and are without legal protection are those who live in illegal settlements consisting

> especially of slums, clandestine subdivisions, or isolated dwellings. People without legal protection usually are those who are in an illegal situation. For the most part, authorities do not repress the poor when their illegal situation is limited to lack of a property title, allowing them to remain in place while expropriating the property or resorting to concessions for the use of urban land, among other urbanistic instruments. Eviction occurs more often in the case of invasion of urban areas and housing staged by homeless people or by popular pro-housing movements. Eviction occurs more often in critical situations to restore property ownership. (Brazil 2001: para. 509)

In this quotation, Brazil raised and linked several subjects in only a few sentences: eviction, legal protection, expropriation, illegal settlements, slums, clandestine subdivisions, isolated dwellings, homelessness, property title, and property ownership. Although Brazil was using the term illegal (instead of informal), the country demonstrated a remarkably new view on informal housing structures. Brazil's quotation also stands for a new trend. Brazil (and other states) started to respect and to protect informal, or even illegal, housing and working structures. They no longer repressed the poor when their illegal situation was limited to the lack of a property title. However, the country seemed to differentiate between the poor without legal protection and homeless people or pro-housing movements that invaded urban areas. The Dominican Republic even spoke of the informal sector as a recognized major group in the field of housing:

> In regard to the quantitative deficit, there is an estimated need to build about 357,500 new housing units, in an effort to address this acute human need, involving two major groups:
>
> The formal sector, which covers the public sector through its various government and municipal institutions, and the private sector, which encompasses companies, organizations and individuals with established economic, technical and legal capacities. This sector covers only 25 per cent of annual production of new housing;
>
> The informal sector encompasses all the initiatives of individuals and community groups who are not covered within the legal and technical

> framework governing the housing sector, and yet it covers 75 per cent of annual production of housing. (Dominican Republic 2008: para. 197)

The quotation shows one reason that might explain why states, such as the Dominican Republic, started to recognize the informal sector. This sector covered three-quarters of the country's housing production. In sum, the third period showed a sort of changing view about the relationship between illegal and informal housing. Even if the lexeme illegal was mentioned more than the lexeme informal, as the relation was 91% (illegal) to 9% (informal)[4] in the second period, the significance of the lexeme informal rose up to 39%[5] (to 61% for illegal) in the third period. This new recognition of informal structures was often based on an economical point of view. The states recognized the interdependence between informal and formal structures.

> Official statistics estimate that 42 per cent of all jobs are in the informal economy. This affects all economic sectors to varying degrees (catering, textiles, tourism, domestic work, construction and transport). Although it has been ignored in the past, the informal economy is now at the centre of concern. Studies have shown that the informal production system is closely linked to the mainstream economy. This interdependence has led Djibouti to adapt its macroeconomic and microeconomic frameworks and to take 'appropriate measures to guide the informal sector toward the modern economy'. (Djibouti 2010: para. 49)

Djibouti also acknowledged that the informal economy has been ignored in the past. Morocco (2013: para 63) reported that any attempt to upgrade the national production system would fail until the country considers the important role played by the informal sector in employment and wealth creation. Many states also started to adjust to non-formal housing, albeit carefully. This new trend also spread into the

---

[4] N = All terms illegal, informal, illegality, informality mentioned in the paragraphs on inadequate housing of the States Parties reports from 1990 to 1999.

[5] N = All terms illegal, informal, illegality, informality mentioned in the paragraphs on inadequate housing of the States Parties reports from 2000 to 2015.

Committee's Reporting Guidelines: While the Committee asked the states to provide information concerning illegal settlements in its second Reporting Guidelines a few times (CESCR 1991b), the term illegal did not come up at all in the third Reporting Guidelines (CESCR 2009).

The rising recognition of informal housing structures can also be proven by more attention given to the land titling approach. More than one-fifth of the reporting states mentioned land titling as a measure for Spaces of Inadequate Housing (e.g., Albania 2010: para. 309; Guyana 2012: para. 324; Nicaragua 2007: para. 782–786 and para. 808–810; the Philippines 2006: para. 637–696; United Republic of Tanzania 2009: para. 107). The next quotation, however, shows some uncertainty:

> The legislative framework and the programmes for the temporary 'certification' of illegal structures and, at later stages, for their integration into a new official city plan, have been purposefully designed in such a manner that the negative impacts for those living in illegal housing are minimized. Thus, the actual cases of partial or total demolition or expropriation of property due to unsolvable planning problems have been kept to a minimum. Moreover, special exemptions have been provided in planning law in order to facilitate the expansion, improvement or further development of housing in illegal properties and expedite the process of 'legalization'. Needless to say, this has often negative impact from the point of good planning and quality of the built environment, but is considered necessary due to the particular social composition of former illegal housing areas. In all, the regime for the 'integration' and 'legalization' of illegal housing can be reasonably considered rather beneficial for past illegal settlers both from the point of housing and from the point of property gains. (Greece 2002: para. 323)

The new policy described contradicted the country's planning philosophy, but seemed to be unavoidable 'due to the social composition of former illegal housing areas.' It is not a coincidence that this uncertainty arose in the report from a European country. European states were overall less likely to report on land titling. In contrast, countries from Latin America and the Caribbean (e.g., Argentina 2009: para.

669; Uruguay 2009: para. 228–229; see also Kolocek 2012) reported on the titling policy more often and with much more self-evidence:

> The basic purpose of the 'El Salvador, Land of Owners' programme is to provide legal certainty by awarding title on a [sic] individual basis and swiftly, efficiently and safely to the plots occupied by thousands of poor families in marginalized areas, 'pirate' settlements, unauthorized communities, etc., and entering the titles in the Public Land Register. Title has been awarded to 1,629 families living in 37 communities in eight of the country's departments. On the basis that an average family has five members, this programme has helped 8,145 Salvadorans to become authentic and legitimate owners. (El Salvador 2004: para. 663)

Surprisingly, the World Bank, which was regarded and criticized as de Soto's main partner in the scholarly discourse, hardly came up in the states' descriptions of their land titling policies. In the case of housing, if at all, the World Bank was more likely mentioned when the states talked about the improvement of infrastructure or technical assistance and funding (e.g., Georgia 2001: para. 177; Kenya 2006: para. 31), usually as one actor of many other supranational actors:

> The Government has also embarked on urban modernisation projects with the assistance of the World Bank and HABITAT, a United Nations agency for human settlement. The project aims at modernising old, unplanned and squatter areas by constructing different types of infrastructures like roads, sanitation facilities and social facilities like community health centres. (United Republic of Tanzania: 2009: para. 102)

However, land titling was not only mentioned in an urban housing context, but also in rural areas. The states then talked about land titling and regularization of agricultural land in the context of the human right to food (e.g., Costa Rica 2006: para. 784–787; Dominican Republic 2008: para. 231–232, Guatemala 2011: para. 242–244). The next quotation will once again demonstrate the changing perception of informal housing and the high density of reported measures, involved actors, and forms of inadequate housing:

> Right to shelter is recognized as an integral part of the fundamental right to life under the Constitution of India. The Supreme Court has through many judgements stressed the importance and spelt out the content of this right. [...]
>
> Housing is the responsibility of the State government under the Constitutional scheme in the country. The Union government however has been taking initiatives and directing the programmes in that sector through the process of planning, by formulating policies, establishing institutions for financial and technical support, implementing social housing development schemes, alongside amending laws and implementing fiscal measures to provide incentives for investing in the housing sector.
>
> In the background of Global Shelter strategy, the National Housing Policy 1988 was formulated to eradicate houseless [sic], to improve housing conditions of the inadequately housed and to provide a minimum level of basic services and amenities to all. The policy envisaged a major shift in the Government's role to act more as a facilitator then [sic] as a provider.
>
> India is a member of United Nations Center for Human Settlement Habitat (UNCHS) and attended the Istanbul + 5 UNCHS (Habitat) Conference. India adopted the Habitat Agenda and has decided to locate a National Urban Observation in the TCPO. The country has submitted a report on the progress of implementation of the Habitat Agenda (1996–2000) in 2001.
>
> After a thorough review, the National Housing and Habitat Policy 1998 (NH & HP) was evolved stressing the issues of sustainable development, infrastructure development and strong public private partnerships for shelter delivery. The NH & HP 1998 is under review & revision due to major changes in habitat & human settlement issues since 1998 and recommendation emerged in the conference of Housing Ministers' of all States in November, 2004. The revised draft policy is under consideration and is likely to be finalized shortly. (India 2006: para 408–412)

The country's manner of discussing housing illustrates a new perspective on the human right to housing for a couple of reasons. First, India stressed the constitutional recognition of housing. This constitutes a difference—one could say progress toward social citizenship—from their reports in earlier periods. In earlier years, India (1983: para. 125) had still reported that the right to housing was not provided for as a constitutional or statutory right. Second, the country's view on housing became more global. India regarded it as necessary to mention its membership in the UNCHS and its attendance of the + 5 UNCHS Habitat conference. Third, although India reported housing as 'the responsibility of the State government', other actors besides governmental actors were mentioned. The state regarded itself as the responsible actor, but noted that it acted more as a facilitator than as a provider. Housing was related to other policy goals and regarded as a planning issue, linked with sustainable development. The language had increasingly become a language of rights, and people affected by inadequate housing were now regarded as citizens:

> The right to adequate housing must be considered within the urban and the rural contexts. The framework of the national policy of urban development adopted by the Brazilian government falls under the wider perspective of the **right to the city.** This right, which is ensured under Arts. 182 and 183 of the Federal Constitution and regulated by the Urban Statute (Law 10257/2001), consists of the right to proper housing, full environmental sanitation, urban mobility for all, urban, legalized land, and a territorial order that meets the needs of all citizens. The establishment of the Ministry of Cities in 2003 sought to consolidate the right to the city at the institutional level, gathering under its umbrella the dwelling areas, environmental sanitation, urban mobility, and territorial planning, and formulating an integrated urban development policy. The recognition of the city's and of property's social function, expressed in the right to the city, is of particular relevance for Brazilian cities in view of the weight and extension of precarious settlements. (Brazil 2007: para. 352, author's emphasis)

The state underpinned the significance of the right to adequate housing as part of 'the right to the city.' Although Brazil explained the right to adequate housing 'within the urban and the rural contexts', the state demonstrated in this quotation that the focal point was placed on urban rather than on rural housing. Here, too, the constitutional roots of housing were mentioned. Moreover, Brazil explained what the state understood under the right to the city. It did not only encompass the right to proper housing, but also the right to full environmental sanitation, urban mobility for all, urban, legalized land, and a territorial order that meets the needs of all citizens. The state stressed the relevance of the recognition of the city's and of the property's function in the context of precarious settlements. One could conclude that Brazil's reporting about the right to the city is a description of a policy for people affected by inadequate housing under the umbrella of social citizenship. Citizenship became more prominent both as a policy goal and as a term. Portugal (2011: para. 115–118) and Tajikistan (2011: 179) explicitly mentioned citizenship for refugees, and Greece (2012: 191) talked about full citizenship rights for Greek Roma in the context of the right to housing. Another way of talking about housing in the face of social citizenship can be proven through the states' new focus on integration goals. China reported an 'Action Plan' that responded to many rights and needs of homeless persons in Hong Kong, a good example of what a housing policy could look like that is both in the face of social citizenship and under the umbrella of de-commodification of land use:

> The Action Plan was completed in March 2004 and was found to be effective in helping street sleepers give up street sleeping. In the light of the effectiveness of the Action Plan, the SWD regularised the services for street sleepers by setting up three Integrated Service Teams for Street Sleepers in April 2004. The Integrated Service Teams which were operated by three NGOs provide a package of tailor-made and one-stop services, including day and late-night outreaching visits, emergency shelter and short-term hostel placement, counselling, employment guidance, personal care (e.g. bathing, hair-cutting and meal service), emergency relief fund, arrangement of long-term accommodation, aftercare service and service referrals. The Integrated Service Teams also collaborate with other NGOs such as the Society for the Aid and Rehabilitation of Drug Abusers and the Society for Rehabilitation and Crime Prevention, Hong

> Kong, to ensure that specialised services are effectively provided to street sleepers with special needs, e.g. drug addicts and ex-offenders. (China, Hong Kong 2010: para. 11.70)[6]

The state described in detail the importance of emergency shelters and short-term hostel placements, but also showed that personal care issues such as bathing, hair-cutting, and meal service were important, too. The special needs of drug abusers and ex-offenders were recognized as well. What now seems to sound self-evident for many social workers and scientific experts on homelessness, must here be regarded in a historical context. Such a policy was unlikely to be found in a State Party report submitted between 1990 and 1999 and did not come up in reports submitted in the 1970s or 1980s.

Another new trend came up in terms of considering the responsibilities. States started to explain in detail that the governments or other governmental actors started to cooperate with other actors, mostly from the group of NGO+ (e.g., New Zealand 2001: para. 377). They described how new actors, initiatives (the United Kingdom 2001: para. 11.141), programs (Paraguay 2006: para. 436), networks, or associations, emerged:

> Shelters are established by non-governmental non-profit organizations, churches and municipalities. Funds for operating shelters usually come from several sources (domestic and foreign): from the budget of the Ministry of Labour and Social Affairs, district offices and municipalities, foundations, and people's own resources (e.g., fees from clients). In 1992 the Association of Representatives of Shelter Operators was created. This civic association connects various kinds of shelters into a network, permits the exchange of information and experience, and creates prerequisites for coordination and for dealing with State and legislative bodies at the central and local levels and other institutions. It thus builds information connections which permit ongoing monitoring of not only the bed capacity of individual facilities and their occupancy rate, but also client

[6] China was numbering the here quoted report in a different manner than most other countries did. '11.70' means paragraph 70 in the section concerning Article 11 ICESCR. The United Kingdom (2001) did this as well.

> turnover. In mid-1999, 59 shelters were members of the association. (Czech Republic 2000: para. 424)

In the case of the Czech Republic, housing for homeless people was offered on the local level through nongovernmental, nonprofit organizations, churches, and municipalities, and paid from several national and supranational sources. The human right to housing was no longer an issue that was claimed by large-scale construction programs, but an issue for cooperation and coordination between several actors on different levels. This can also be proven numerically. In the second period, the range of the mentioned actors in the states' discussion of inadequate housing had grown. This trend continued and was even strengthened in the third period. Once again, governmental actors dominated, but the number of actors from the other six groups was still increasing. On average, each state that reported inadequate housing mentioned the involvement of 3.5 (of seven) groups of actors. In more detail: 102 of the 108 states (94%) that reported inadequate housing talked about an involvement of governmental actors, 56 states (52%) mentioned NGO+, 59 states (55%) mentioned local actors and 48 states (44%) mentioned market actors. In particular, the number of states mentioning supranational actors (45 states, 42%) as well as self-help/family (30 states, 28%) has significantly increased in relation to the former period. Adequate housing became an important issue for both local and global actors. The third period shows a renaissance of the group family/self-help. The family or self-help played a significant role in the first period in the context of rural housing. This group then became less important in the second period when housing problems and policies became more concentrated in urban areas. In the third period, states began to recognize the importance of the family or self-help in the urban context.

The Covenant explicitly obliges the states for international assistance and cooperation by realizing ESC rights (Article 2, para. 1, ICESCR). The Committee has also repeated this obligation in many paragraphs of its second Reporting Guidelines, including the housing section (CESCR 1991b: 15). In the third period, the states started to take this reporting obligation more seriously. As shown, the states' manner of talking about housing became more global.

More than 80% of the reporting States Parties mentioned the Millennium Development Goals, the United Nations Development Programme, or other international programs and agreements. The World Bank, for instance, is a typical supranational actor and one of the most influential organizations in the field of global social policy. Additionally, as discussed earlier, the World Bank was highly involved in land titling discourses. The Bank supported social policies in many states worldwide for several decades (Deacon 2007). The incidence of the World Bank in the States Parties reports continuously increased. The Bank was mentioned more often, however, when the states reported the implementation of other ESC rights. In the last period, only a few states talked about an involvement of the World Bank in their housing sections (e.g., Afghanistan 2007: para 93–94; the Democratic Republic of the Congo 2007: para. 187). The involvement of the World Bank in land titling policies was hardly reported.

States began to regard homeless people through the lens of (in-) equality (e.g., Finland 2005: para. 80–81; Poland 2001: para. 372 and 2007: para. 31; the United Kingdom 2007: para. 155 and 319). Sometimes, they talked about them even as rights holders (e.g., Canada 2004: para. 86; India 2006: para. 408). At the same time, states connected homeless people with health, mental illness, and drug abuse issues (e.g., Italy 2003: para. 196; Latvia 2005: para. 452; the Netherlands 2005: para. 345; Poland 2007: para. 692; Sweden 2006: para. 377–378). Accordingly, they were more frequently considered as members of the society—as citizens—although they were still sometimes treated as vulnerable victims.

By considering the measures to respond to homelessness, a new trend arose, especially in the reporting of Western states. Night shelters, emergency shelters, and hostels were frequently the center of attention when States Parties talked about homelessness (e.g., Canada 2004: para. 339–340; Estonia 2001: 542; Spain 2002: para. 392–393). The idea of the sheltering policy was that homeless people should not sleep outside. Additionally, the states reported social support and reintegration programs offered in these shelters (e.g., Czech Republic 2000: para. 423–425; Hungary 2005: para. 406–415) although some researchers questioned the qualities of such shelters (e.g., Ajayi et al. 2011; Busch-

Geertsema & Sahlin 2007). It seems that emergency shelters or night shelters could, under certain circumstances, be counted as Spaces of Inadequate Housing. Sometimes, the states mentioned this risk in a self-critical manner, like Hungary (2005: para. 406) or France did:

> Examination of the department plans for emergency accommodation reveals that in most departments capacity either meets or exceeds the statutory objective. However, it is also apparent that sometimes the accommodation is unsuitable (use of dormitories, inadequate sanitation, dilapidated buildings, poor geographical location, failure to adapt to the new groups affected, or to young people and families), that there is a need for all-year-round facilities, and that the accommodation must be organized in series, so as to allow clients to progress in step-wise fashion to independent accommodation. Aside from the specialized facilities and the hostels reserved for particular groups, it seems that temporary accommodation is still lacking in certain areas. (France 2000: para. 539)

As discussed, in the third period, the states' attention to both main forms of inadequate housing rose, and at the same time changed. This also applies to the vulnerable and disadvantaged groups. In this third period, 40% of the 118 reporting states mentioned asylum seekers in their reports, 25% mentioned Roma, 47% mentioned indigenous people and more than three-quarters of the reporting states talked about refugees. 102 States Parties (86%) mentioned at least one of the four groups in general, and 72 states (61%) talked about at least one of the four groups in the context of housing. Again, other rights were often touched upon when these groups came into focus. Refugees received the most attention. Forty-five countries talked about housing for refugees; this is about half of the 88 states that mentioned refugees at all. When states talked about refugees, some states meant returning refugees who had left the country for a while because of armed conflicts (Afghanistan 2007: para. 15; Bosnia and Herzegovina 2004: para. 335–350). Armenia explicitly acknowledged that the country differentiated between exiled Armenians and refugees from other countries such as Iraq, Iran, or Georgia (Armenia 2011: para. 274). Other states reported considerable differences in the housing standards between asylum seekers and other groups in need. Germany, for instance, acknowledged that the housing measures for asylum seekers probably did not live up to adequate housing standards:

> [C]ollective reception centres for asylum-seekers do not offer permanent accommodation but merely meet the need for temporary shelter. Consequently, asylum-seekers must accept the inconveniences which typically come with a stay in collective reception centres. This point has been confirmed by German constitutional case law. (Germany 2000: para. 195)

Germany is one of a few countries whose manner of discussion demonstrated that, when it came to housing, asylum seekers were somehow regarded as 'second-class' citizens. Norway also reported that the state differentiated between national citizens and non-nationals:

> In principle, asylum-seekers in Norway are not entitled to their own home. However, the State offers lodging to all asylum-seekers. This follows from the Government's annual budget proposition, which is adopted each year. The asylum-seeker may choose whether or not to accept the offer.
>
> Foreign nationals who have been granted permission to reside in Norway on the basis of an application for asylum and do not manage to find a place to live themselves are offered settlement in a municipality. Municipal authorities decide how many refugees they wish to settle, in cooperation with the State. Negotiating this type of settlement can be time-consuming and difficult. Consequently, many asylum-seekers in Norway live in reception centres for a certain period of time after their residence permit has been granted. (Norway 2004: para. 220 and 221)

Here again, the cooperation of actors on different levels becomes visible, although the quotation demonstrates that the cooperation is more a negotiation about responsibilities for unwanted problem groups. Municipalities and the state could not agree how to deal with refugees; consequently, many asylum seekers still live in reception centers. The Committee took notice of this and demanded equal treatment. In its

Concluding Observations considering a State Party report from Austria, the Committee showed its concern because social assistance benefits provided for asylum seekers were considerably lower than for citizens of the country (CESCR_AUT 2006: para. 15).

Other countries were friendlier: 'Pursuant to the Integration Act, all newly arrived refugees in Denmark must be offered a permanent dwelling within three months. To promote the refugees' integration into Danish society, the refugees are housed across the entire country' (Denmark 2003: para. 320). Denmark's refugee policy does not only show ingredients of global social citizenship. It also demonstrates how the country tried to avoid spatial segregation.

I did not examine the differences in detail between men and women in the reporting on inadequate housing. In the beginning, here and there some states reported on emergency housing for women (e.g., Australia 1980: p. 44), but in the last period, the states, by presenting statistics on homelessness, for instance, began to differentiate between women and men. In these statistics, usually more than 80% of the homeless people were male (e.g., Israel 2001: para. 306; Italy 2003: para. 197). In Cambodia (2008: para. 512), this was also true for street children. A new topic came onto the agenda: women's land rights. African countries started to report on women's access to land ownership (e.g., Kenya 2013: para. 58; Togo 2010: para. 315), rural women (Egypt 2010: para. 381), or housing credits for female employees:

> Burundian women can now be said to have the same rights as men to access to credit from banks. Some female employees and businesswomen are acquiring building plots for houses. This is a considerable advance, since even the central bank, which had long resisted giving mortgages to married women, has finally changed policy. (Burundi 2013: para. 69)

Between 2000 and 2015, the Committee published Concluding Observations addressed to 123 States Parties. From the perspective of housing, these Concluding Observations significantly differed from the Concluding Observations in the second period. The Committee's attention to housing issues intensified considerably. Of the states, 61% were addressed about homelessness; in the case of Spaces of Inadequate

Housing, it was 58%. For one-fifth of the addressed states, neither Spaces of Inadequate Housing nor homelessness were touched upon. The Committee's demonstrated interest in street children grew from 19% in the second period to 27% in the third one. The Committee strengthened its policy to point out the risks of privatization (e.g., CESCR_CZE 2002: para. 19; CESCR_HRV 2001: para. 34; CESCR_ITA 2004: para. 26; CESCR_MAR 2006: para. 26).

> The Committee notes that the 'economy-first' approach adopted by the State party has resulted in a low priority being placed on the protection of economic, social and cultural rights. This has led to the marginalization of certain groups in society in such matters as housing, social welfare and health care. (CESCR_KOR 2001: para. 11)

While the Committee focused most of its interest in the 1990s on housing issues and the two General Comments, in the 2000s, the Committee reminded the states of their duties more vehemently. The Committee addressed 73% of the states regarding eviction, most of them with urgent appeals. What was new in the Committee's way of addressing eviction was that evicted minority groups were taken into consideration, for instance Roma (e.g., CESCR_BIH 2006: para. 25; CESCR_GRC 2004: para. 21; CESCR_MKD 2008: para. 23; CESCR_SRB 2014: para. 30) or indigenous people (e.g., CESCR_BOL 2001: para. 21; CESCR_BRA 2003: para. 58).

> The Committee notes with deep concern the large number of forced evictions of peasant and indigenous families, particularly in the communities of Tetagua Guarani, Primero de Marzo, Maria Antonia and Tekojoja, who had been occupying the land, and the reports received that the National Police used excessive force in carrying out those evictions, by burning and destroying housing, crops, property and animals.
>
> The Committee notes with concern that some 45 per cent of indigenous people do not hold legal title to their ancestral lands and are thus exposed to forced eviction. (CESCR_PRY 2008: para. 17–18)

The absence of a legal title was here regarded as the reason for forced evictions. The Committee often reminded the States Parties of the guidelines adopted in the two housing-relevant General Comments and used reports of the Special Rapporteur on adequate housing to claim ongoing forced evictions in the countries (e.g., CESCR_AFG 2010: para. 39). In its Concluding Observations for the Democratic Republic of the Congo, the Committee (CESCR_COD 2009: para. 39) also offered an invitation of Special Rapporteurs and other experts. Moreover, the Committee encouraged some states to ratify ILO Conventions (e.g., CESCR_KHM 2009: para. 16; CESCR_MDG 2009: para. 36; CESCR_TCD 2009: para. 38) or recommended international assistance to States Parties:

> The Committee notes with satisfaction that the Government of Georgia is willing to cooperate with various international organizations, such as the United National Development Programme, ILO, World Health Organization, the World Bank and the International Monetary Fund, as well as with regional organizations such as the Organization for Security and Cooperation in Europe, in order to realize the rights set forth in the Covenant, and in particular to address the problem of poverty.
>
> [T]he Committee recommends that the State party ensure that its international human rights obligations are taken fully into account when it enters into technical cooperation and other arrangements with international organizations. Also, the Committee urges the State party to take into account the different approaches taken by various international organizations, such as the human development approach of UNDP. (CESCR_GEO 2000: para. 4 and para. 22)

Poverty was regarded as an issue that should be challenged by the States Parties through cooperation with international organizations. The view on informality continued to change. However, this does not mean that the Committee did not point out potential dangers. In the Concluding Observations concerning a report of Monaco, the Committee was concerned about the informal employment in the hotel and catering and construction industries, particularly the working conditions of the persons employed in these sectors (CESCR_MCO 2014: para. 15).

Moreover, unemployment rates and low wages were regarded as the main problems in the informal economy:

> The Committee is concerned that unemployment remains high in the State party, in particular in rural areas and among young persons, women, indigenous and Afro-Colombian peoples. The Committee is also concerned that the creation of employment opportunities is taking place primarily in the informal economy (60 per cent) with a negative impact on access to social security. The Committee is further concerned about the working conditions in the informal economy and rural areas where wages remain very low (arts. 6, 7). (CESCR_COL 2010: para. 11)

The Committee did not urge the states to prohibit or prevent informal working, but rather advised them to remedy the circumstances:

> The Committee invites the State party to provide in its next periodic report detailed information, including statistical data disaggregated by sex and by rural and urban distribution, on the extent of the informal economy, as well as the State party's policies and protection measures, if any, taken to deal with it. (CESCR_POL 2009: para. 35).

Some States Parties started to include a section in their reports in which they directly replied to the Committee's Concluding Observations to the states' previous reports (e.g., Czech Republic 2010: Annex 1; Mexico 2004: para. 935–1242). Bosnia and Herzegovina's (2010) second periodic report consisted of an introduction and detailed responding to 21 recommendations of the Committee. The reporting system more and more became a sort of discussion:

> In its Concluding Observations, the Committee on Economic, Social and Cultural Rights called upon Georgia to take appropriate measures to create conditions that would allow internally displaced persons to return to their places of origin (para. 29). Unfortunately, we have to report that, given the lack of a political solution to the conflicts alluded to in the initial report, Georgia has been unable to comply with this particular recommendation of the Committee during the reporting period. (Georgia 2001: para. 184)

The discussion between the Committee and the ICESCR member states went from a simple reporting–asking–reporting–asking rhythm into a disputing direction. On the one hand, as seen, the Committee started to use more sources (like the reports of the Rapporteurs or other experts, or UNPD reports) to claim violations of ESC rights. On the other hand, some states started to question the Committee's suggestions and claims. China, for instance, received some urgent appeals from the Committee regarding the bed-space apartments and cage homes in Hong Kong, which were, from the Committee's perspective, an affront to human dignity. Moreover, rooftop structures constituted a grave risk to the life and health of their inhabitants (CESCR_CHN 2001: para. 25). By responding to these critics, the state promised to clear illegal rooftop structures and re-house the occupants as well as the squatters. With respect to bed-space apartments, the state did not accept the Committee's suggestions:

> [W]e appreciate the Committee's concerns but we believe that our approach is the right one. The people who live in bed-space apartments do so largely out of choice for convenience, economic and other personal reasons. There is ample provision in government hostels—which are clean, safe and well managed—to accommodate the relatively small numbers involved. Yet they remain underutilized, probably because their targeted occupants place a higher value on location and convenience than on the standard of their living environment. It would not be practical to plan the provision of hostels or other forms of public accommodation on the basis of where individuals prefer to live, particularly as such preference is invariably for the urban areas where space is already at a premium. We cannot force people to accept public accommodation if they do not want to. Nor do we think it reasonable to abolish a form of private sector accommodation for which there is a small but persistent demand. Thus, our policy has been to enforce an amelioration of conditions within the premises that provide bed-space apartments. Now, all bed-space apartments are regulated under a statutory licensing regime that enforces fire and building safety standards. Meeting those standards has obliged the operators of such premises to reduce the number of bed-spaces per unit, so that the apartments are not as cramped as they used to be. In this way, we have respected the personal choice of those who live in such accommodation, whilst achieving substantive improvements in the

> standard of their living environment and offering them genuine public sector alternatives. (People's Republic of China 2003: para. 592)

Scholars interested in civil and political rights might be surprised that China's key argument was the individual's preferences, 'the personal choice', and that the country could not 'force people to accept public accommodation if they do not want to'. In the next Concluding Observations responding to the just quoted report, the Committee did not react to the state's opinion directly. It only reiterated its concern about the persistence of inadequate housing in the form of cage-homes and bed-space apartments (CESCR_CHN 2005: para. 78). Israel also contradicted the Committee's comments:

> The information received by the Committee regarding ownership claims on Bedouin lands, is inaccurate. The Land Settlements Department was not established to address Bedouin land claims, but was established by the British Mandatory Government during the 1930s to deal with various land claims in the absence of legally-registered rights in mandatory Palestine. Since then, the department has addressed title claims involving all sectors of the population in Israel, including inter alia Bedouins claims. (Israel 2001: para. 359)

In the third period, a trend became more apparent that had begun at the end of the second period. The Committee heavily criticized the consequences of the structural adjustment programs (SAPs) (e.g., CESCR_BOL 2001: para. 9; CESCR_CMR 1999: para. 10), especially for the most disadvantaged groups (e.g., CESCR_ARG 1999: para. 10; CESCR_BRA 2003: para. 16; CESCR_COL 2001: para. 9; CESCR_HND 2001: para. 10).

> The Committee is of the view that the state of emergency that has been in place in Egypt since 1981 limits the scope of implementation of constitutional guarantees for economic, social and cultural rights; that some aspects of structural adjustment programmes and economic liberalization policies introduced by the Government of Egypt, in concert with international financial institutions, have impeded the implementation of the Covenant's provisions, particularly with regard to the most vulnerable

> groups of Egyptian society; and that the persistence of traditional practices and attitudes, deeply entrenched in Egyptian society, with regard to women and children hamper the ability of the Government to protect and promote their economic, social and cultural rights. (CESCR_EGY 2000: para. 10)

As the structural adjustment programs enhanced inequality by affecting the most vulnerable, they contradicted the main principles of the ICESCR. Therefore, the Committee demanded the states consider the effects on this group:

> The Committee calls upon the State party, when negotiating with international financial institutions and implementing structural adjustment programmes and macroeconomic policies affecting foreign debt servicing, integration into the global free market economy, etc., to take into account their effect on the enjoyment of economic, social and cultural rights, in particular for the most vulnerable groups of society. (CESCR_MEX 1999: para. 34)

The quotation is remarkable because it shows that, although the Committee tried to push the States Parties into a global social policy direction, at the same time, the Committee was skeptical of some of the most prominent programs of global social policy. Moreover, the Committee regarded market forces more as a danger than as an opportunity.

All in all, in the third period, inadequate housing was regarded as a problem that required individual solutions. This trend has its roots in the second period and is a key difference to the states' reporting in the 1970s and 1980s when homelessness was only seldom mentioned, and housing was either a rural issue or a problem of affordability that the governments solved through mass construction programs. In the third period, the recognition of the complexity of inadequate housing even increased in relation to the second period. Language that shows ingredients of social citizenship had spread into the way many states discussed both forms of inadequate housing. The states increasingly recognized homeless people and Spaces of Inadequate Housing dwellers as right holders or tried to give the people 'new rights', property rights for instance.

In the case of homelessness, the new perspective is proven by the focus on integration goals, often combined with measures that responded to other needs. Homeless people were regarded as citizens who have rights. This did not only apply to their right to housing, but also to other rights such as the right to food, the right to health, or the right to education. The strategic benefits of the new night shelter policy of mostly Western countries have to be answered for each case individually. It seems that a few states realized to some extent that they could not entirely avoid inadequate housing; they had to deal with it somehow. Perhaps night shelters, emergency shelters, and hostels that tried to focus on social integration were both the first steps out of homelessness and a new form of Spaces of Inadequate Housing at the same time.

In the case of Spaces of Inadequate Housing, the road to social citizenship is more of an economic one. Dwellers of former illegal settlements, slums or informal settlements were often more likely regarded as entrepreneurs who should, through land titling or microcredits, be integrated into the market society. The new 'regime[s] for the "integration" and "legalization" of illegal housing' (Greece 2002: para. 323) were supposed to result in economic win-win situations: Increase social citizenship through market participation. The Committee, on the other hand, was particularly skeptical with respect to market forces.

In the language of the Maastricht Guidelines (1997), one could say that the states started to regard their new primary task in respecting housing rights and protecting people from a violation, while the fulfill element increasingly became a task for other actors: market and supranational actors in the case of Spaces of Inadequate Housing, NGO+ in the case of homelessness, and supranational actors in the case of street children.

However, the states' way of talking about inadequate housing showed more and more ingredients of both social citizenship and a globalization of social policy. The road to social citizenship was mostly an economic one. The states showed only traces of de-commodification of land use. States from Europe showed traces of de-commodification of land use when they claimed that their night shelters responded to more rights and needs than merely to the need to have a roof over one's head for a few nights.

The next chapter will regard, as a whole, the States Parties reports submitted between 1977 and 2015. It will point out the main differences between the reporting on homelessness and Spaces of Inadequate Housing and put a finger on the different foci in the UN regions of Africa, the Americas, Asia, Europe, and Oceania.

6

Comparative Discussion: Interpretations of the Human Right to Housing

Chapter 5 concentrated on the narrative of the discourse and discussed in detail how the states' view on inadequate housing has changed between 1977 and 2015. This chapter focuses on how the UN regions and the states differ in their reporting on inadequate housing with respect to the theoretical framework presented in Chapters 3 and 4. It starts with a statistical overview of the global distribution of certain housing aspects.

6.1 Housing in Numbers: The Global Distribution of Housing Aspects

This section gives an overview of the contents of the research data as a whole. It addresses the following questions: Which states mentioned homelessness and Spaces of Inadequate Housing at all in their reports? How much attention did the states pay to inadequate housing? Which vulnerable and disadvantaged groups dominated in the states' talk about inadequate housing? Which actors appeared in the paragraphs concerning inadequate housing? I show the main differences and overlaps between the UN regions

M. Kolocek, *The Human Right to Housing in the Face of Land Policy and Social Citizenship*, DOI 10.1007/978-3-319-53489-3_6

and present the main contents of the Committee's Concluding Observations. The chapter takes almost all States Parties reports and Concluding Observations into consideration. Only states that do not exist anymore, such as Czechoslovakia, the German Democratic Republic, and Yugoslavia, have not been taken into account.

6.1.1 Inadequate Housing

To give an overview of the states' notice of inadequate housing in a global context, I first measured the attention paid to the main category, inadequate housing, without differentiating between its main forms. The level of attention shows how seriously the states took inadequate housing as a problem (Table 6.1). I differentiated between five levels of attention (including no attention).

In sum, 129 States Parties submitted between one and six full reports. Diffusion Map 6.1 (p. 169) shows that 120 out of 129 countries (93%) reported on inadequate housing, albeit with different levels of attention. States from Africa and Asia only seldom paid a lot of attention to inadequate housing, while states from Europe and the Americas often reported on inadequate housing intensively. The following nine states (7%) neither

Table 6.1 Level of attention to inadequate housing

Level of attention	In detail
Very high	The State Party reported a lot and provided very detailed information on both main forms of IH homelessness and Spaces of Inadequate Housing. It spent many paragraphs on the subject and reported statistical information, causes, and measures to deal with IH
High	The State Party reported on both main forms, but often only reported one of two forms with a very high level of attention
Medium	The State Party mentioned IH with some paragraphs, but did not go into detail. A few States Parties only focused on one main form of IH and neglected the other
Low	The State Party mentioned IH only once or twice with less than three (mostly short) paragraphs
No attention	The State Party did not mention IH

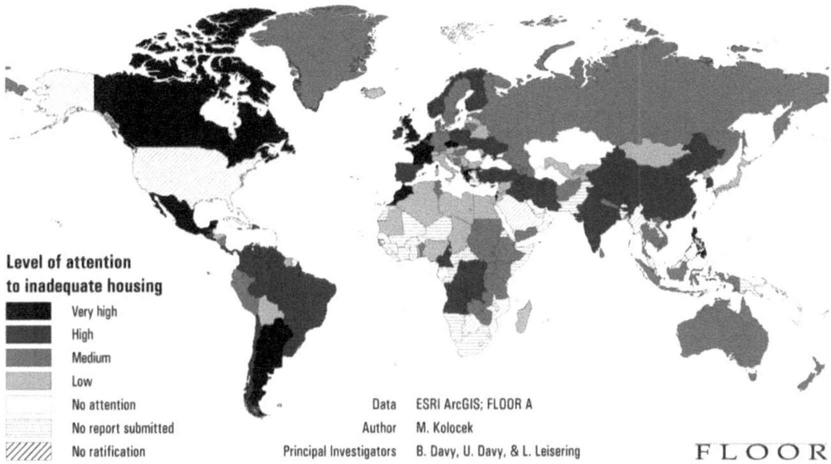

Diffusion Map 6.1 ICESCR mebmer states' attention to inadequate housing

mentioned homelessness nor did they mention Spaces of Inadequate Housing with any word: Barbados, Gabon, Kazakhstan, Kuwait, Lebanon, Romania, San Marino, Senegal, and Turkmenistan. These states will be discussed in more detail in Section 6.2.

As inadequate housing is separated into the two main forms homelessness and Spaces of Inadequate Housing, I asked how many states mentioned these forms at all. Of the 129 reporting States Parties, 107 (83%) mentioned homelessness in at least one of the reports. This also means, however, that 22 states (17%) did not speak of homelessness with a single word. Of the 107 states reporting on homelessness, 68 (64%) talked about street children.

Table 6.2 (p. 170) shows homelessness as an important issue in every UN region, but a mention of homelessness in African States Parties reports was less likely than in the reports of States Parties from the Americas, Asia, Europe, and Oceania. Only about a third of the states from Europe (38%) and Oceania (33%) reporting on homelessness talked about street children, while African states and States from the Americas gave proportionately more attention to this form of homelessness. Of 20 African states reporting on homelessness, 16 (80%) talked about street children; in the Americas, it was 19 out of 21

Table 6.2 ICESCR States Parties mentioning homelessness

UN region	Submitting States Parties	States Parties mentioning homelessness	States Parties mentioning street children
Africa	28	20 (71%)[1]	16 (57%; 80%)[2]
Americas	24	21 (88%)	19 (79%; 90%)
Asia	32	26 (81%)	18 (56%; 69%)
Europe	42	37 (88%)	14 (33%; 38%)
Oceania	3	3 (100%)	1 (33%; 33%)
Sum (average)	129	107 (83%)	68 (53%; 64%)

Source: Database FLOOR A, SPR. Own analysis.

[1] Percentage based on submitting States Parties.

[2] First information in the brackets: percentage based on submitting States Parties; second information: percentage based on the States Parties mentioning homelessness.

(90%). Some countries only mentioned street children when they brought up the topic of homelessness, especially states from Africa (Kolocek 2012).

Spaces of Inadequate Housing is the second code of inadequate housing. Out of 129 States Parties, 104 (81%) mentioned Spaces of Inadequate Housing in their reports (Table 6.3, p. 171). Asia stands out with only 72%. The most frequently mentioned terms for Spaces of Inadequate Housing were illegal settlements, slums, or squatter settlements (including squatting) (Table 6.3). Slums were one of the three most frequently mentioned types in four of five UN regions. The last column in the table demonstrates that, despite the increasing attention to informal housing, in sum, inadequate housing was closely linked with something understood to be illegal.

The numbers show that homelessness was the more prominent form of inadequate housing in Europe and Asia. In the Americas and Africa, Spaces of Inadequate Housing were a little bit more likely mentioned. In total, both forms were reported on a fairly equal level. Out of 129 countries, 91 (71%) reported both homelessness and Spaces of Inadequate Housing (not illustrated in the tables). This means that, if a country reported one of the two main forms of inadequate housing, it likely also reported the other one. Asian states rank last in terms of reporting Spaces of Inadequate Housing and African states rank last in reporting on homelessness.

Table 6.3 ICESCR States Parties mentioning Spaces of Inadequate Housing

UN region	Submitting States Parties	States Parties mentioning Spaces of Inadequate Housing	Three most mentioned terms of Spaces of Inadequate Housing
Africa	28	23 (82%)[1]	Slums, squatter settlements, shanty towns
Americas	24	22 (92%)	Illegal settlements, slums, informal settlements
Asia	32	23 (72%)	Slums, squatter settlements/squatting, illegal settlements
Europe	42	33 (79%)	Illegal settlements, slums, squatter settlements/ squatting
Oceania	3	3 (100%)	Squatter settlements/ squatting, shacks
In sum (Average)	129	104 (81%)	Illegal settlements, slums, squatter settlements/ squatting

Source: Database FLOOR A, SPR. Own Analysis.
[1] Percentage based on submitting States Parties.

6.1.2 Responding Actors

Table 6.4 (p. 172) shows the number of states that mentioned the specific actors when they reported inadequate housing. The table differentiates between homelessness and Spaces of Inadequate Housing and illustrates the incidence of actors in the paragraphs concerning street children (which is a sub-code of homelessness). Governmental actors were mentioned by 87 of the 107 states reporting homelessness (81%), 55 of the 68 states reporting street children (81%) and in the Spaces of Inadequate Housing paragraphs for 91 of 104 states (88%). The high incidence of governmental actors indicates that, habitually, the states reported their measures and policies to respond to inadequate housing. Actors from this group appeared in most paragraphs about inadequate housing and if not, either local actors came up, or no actor was mentioned.[1] The group family/self-help

[1] This applies mostly to States Parties that reported inadequate housing with only a few words.

Table 6.4 Mentioned actors in the context of inadequate housing

States Parties reporting	Homelessness 107	Street children 68	SPIH 104
Governmental actors	87 (81%)[1]	55 (81%)	91 (88%)
Local actors	34 (38%)	13 (19%)	52 (50%)
Family/self-help	11 (15%)	7 (10%)	28 (27%)
Market actors	28 (26%)	7 (10%)	45 (43%)
NGO+	56 (52%)	32 (47%)	38 (37%)
Supranational actors	29 (27%)	21 (31%)	41 (39%)
Other actors	36 (34%)	17 (25%)	33 (32%)
Groups of actors (on average)	**2.7**	**2.2**	**3.2**

Source: Database FLOOR A, SPR. Own analysis.
[1] Percentage based on reporting States Parties.

played a marginal role, especially when states reported homelessness and street children. Local actors were more frequently mentioned when states reported Spaces of Inadequate Housing (50%) and only played a marginal role in the paragraphs on street children (19%). For market actors, 43% of the 104 states mentioning Spaces of Inadequate Housing reported that these actors were involved in the creation of housing policies. With respect to homelessness, it was only one-fourth (26%), and only every one-tenth of the states reporting on street children mentioned market actors in the relevant paragraphs. Supranational actors were mentioned more in the paragraphs about homelessness when the states also reported on street children (31%). However, again, the paragraphs on Spaces of Inadequate Housing showed the highest rate (39%). NGO+ came up more frequently in the sections concerning homelessness (52%) and street children (47%). NGO+ is the only group whose mentions are considerably higher in the homelessness paragraphs than in the Spaces of Inadequate Housing paragraphs. The relation between homelessness and Spaces of Inadequate Housing with respect to other actors (the last group in the table) is similar.

Table 6.4 gives insight into the differences between homelessness and Spaces of Inadequate Housing. When States Parties reported Spaces of Inadequate Housing, many different actors were often involved. In contrast, the diversity of actors in the homelessness paragraphs was much lower. States that reported homelessness reported on an average 2.7 (of seven) groups of actors in the relevant paragraphs, whereas in the paragraphs about street children the rate was lower (2.2). When states reported Spaces of Inadequate Housing, they mentioned 3.2 groups of actors on average. Besides the group of other actors, NGO+ is the only actor that is mentioned proportionately more often when states reported homelessness. From the perspective of the states, homelessness was a task for governmental actors, local actors, and NGO+. Market actors, as well as supranational actors, played a significantly more important role when states reported Spaces of Inadequate Housing. Table 6.4 furthermore underpins the unusual position of street children. Street children were less frequently a subject for the efforts of local actors. Instead, an involvement of supranational actors was mentioned more often. Spaces of Inadequate Housing and street children seem to be more of a global social policy issue than adult homeless people.

The incidence of mentioned actors in the States Parties reports also differs between the UN regions. States Parties from all five considered regions have governmental actors dominating the sections concerning housing. Figure 6.1 shows the importance of a certain group of actors in relation to the other six groups. When States Parties mentioned an actor in the context of the human right to housing, the likelihood that it was an actor from the group of governmental actors was 20% or more in every UN region. In African countries, the likelihood was even more than 30%. Considering the second most frequently mentioned actors, the UN regions differed from each other. In States Parties reports of countries from Europe and Asia, NGO+ were prominent, while the group family/self-help only played a marginal role, but was more prominent in states from the Americas. In States Parties reports from African countries, market actors were in second place. The UN region Oceania includes only three ICESCR member states: Australia, New Zealand, and the Solomon Islands. In these three countries, the four groups of local actors, NGO+s, market actors, and other actors were represented equally, while only one country (Australia 1980) mentioned family/self-help in the housing sections; supranational actors did not play any role.

Figure 6.1 does not illustrate information concerning the density of actors in the states' reporting. This information is interesting as the UN regions differ very much. On average, a state from the Americas talked about 4.8 groups of actors (of seven) involved in policies responding

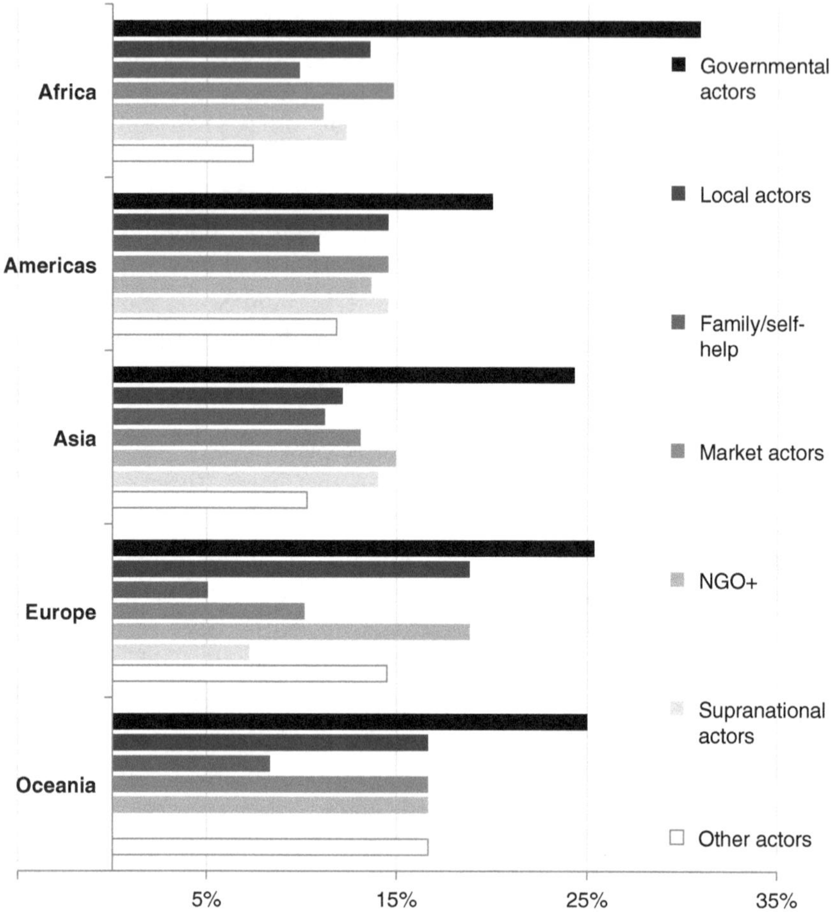

Fig. 6.1 Actors in the context of inadequate housing

Source: Database FLOOR A, SPR. Own analysis. 100 % = number of States Parties (from one of the five UN regions) that mentioned Governmental actors + number of States Parties that mentioned Local actors + number of States Parties that mentioned family/self-help and so on.

to inadequate housing. In contrast, states from the other UN regions showed a significantly lower rate. In the reports of the states from other UN regions, the rate was between 3.1 (Africa) and 4.0 (Oceania). Countries from Europe reported on average 3.5 groups of actors and countries from Asia reported 3.8. All rates are significantly higher than the rates for homelessness, Spaces of Inadequate Housing or street children because the actors for both main forms of inadequate housing were counted. When a state, for instance, mentioned governmental actors, local actors, and NGO+ in its homelessness reporting, and governmental actors and supranational actors in the paragraphs concerning Spaces of Inadequate Housing, it is considered as mentioning four actor groups total.

6.1.3 Disadvantaged and Vulnerable Groups

The quantity of the mentioned vulnerable and disadvantaged groups in the housing sections of the States Parties reports seems to be endless. For that reason, I concentrated on four groups that I discuss in the following paragraphs.

Table 6.5 illustrates that 96 of 129 states (74%) reported refugees and 65 states (50%) mentioned indigenous people. The rates of reporting asylum seekers (38%) and Roma (25%) are much lower. Talking about one (or more) of the four minority groups did not automatically mean that the states also talked about them with respect to housing. Only 50 of the 96 refugees mentioning states (52%) reported them in the context of housing. This means that almost half of the refugee

Table 6.5 Mentioned disadvantaged and vulnerable groups

States Parties mentioning the group	
Refugees	**96** (74%)
Asylum seekers	**49** (38%)
Roma	**32** (25%)
Indigenous people	**65** (50%)
States Parties mentioning the group in the context of housing	
Refugees	**50** (39%; *52%*)[1]
Asylum seekers	**24** (19%; *48%*)
Roma	**20** (16%; *63%*)
Indigenous people	**31** (24%; *48%*)

Source: Database FLOOR A, SPR. Own analysis.

[1] First information in the brackets: percentage based on 129 States Parties. Second information: percentage based on the States Parties mentioning the minority group (96, 49, 32, or 65).

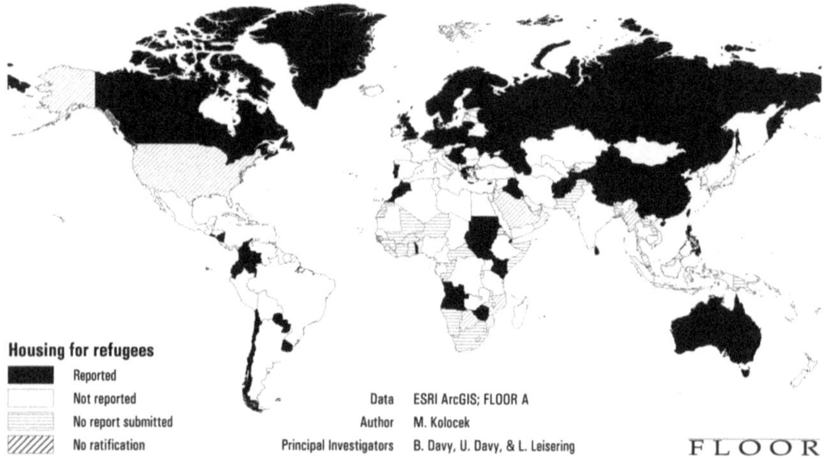

Diffusion Map 6.2 ICESCR member states' attention to refugees in the context of housing

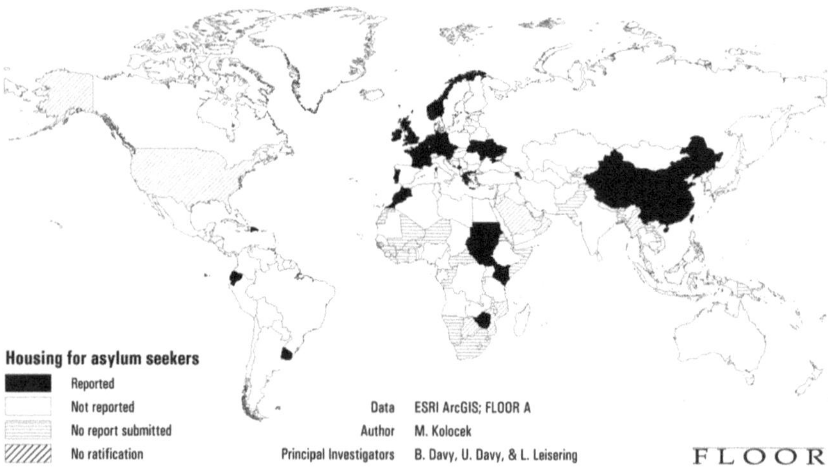

Diffusion Map 6.3 ICESCR member states' attention to asylum seekers in the context of housing

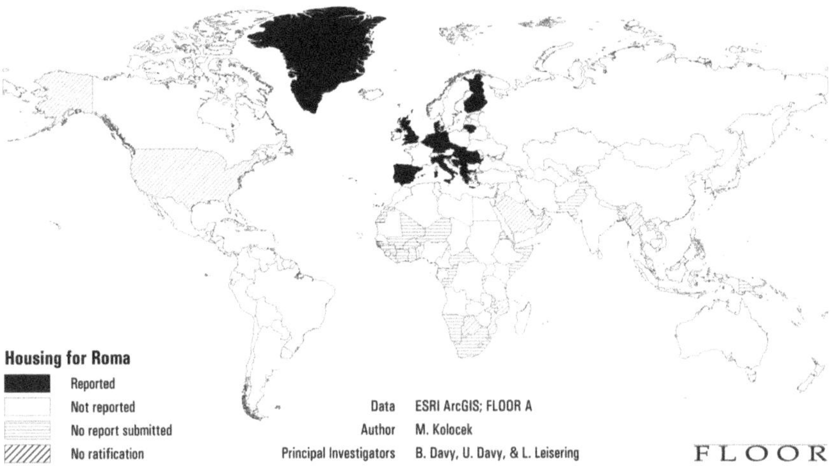

Diffusion Map 6.4 ICESCR member states' attention to Roma in the context of housing

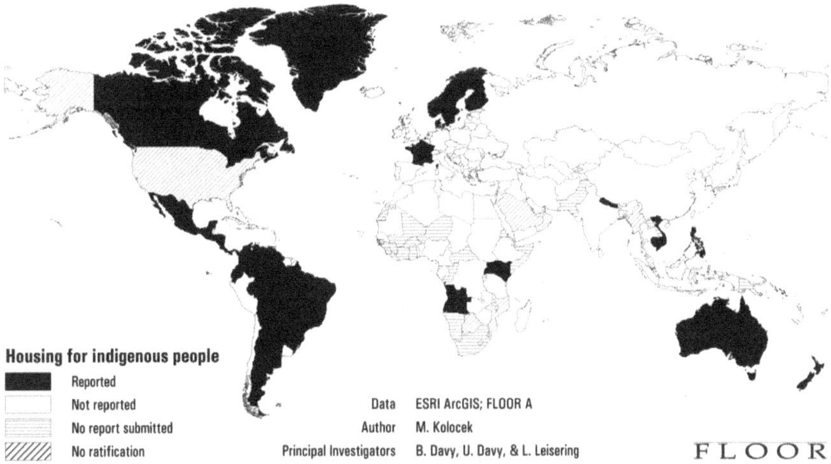

Diffusion Map 6.5 ICESCR member states' attention to indigenous people in the context of housing

reporting states did not touch housing issues in this context. Even lower was the likelihood that states linked asylum seekers (48%) or indigenous people (48%) with housing. These groups were often regarded as important to be discussed in the context of other ESC rights, but not with respect to the right to housing. In contrast to that, Roma were only mentioned by 32 of 129 states (25%), but 20 of these 32 states (63%) talked about this group in the context of housing (Diffusion Maps 6.2, 6.3, 6.4 and 6.5).

The maps illustrate the global distribution of the states' notice to the four vulnerable and disadvantaged groups in the context of housing. No state discussed all four groups with respect to housing, and 12 states—Belgium, Czech Republic, Denmark, Ecuador, Finland, Germany, Greece, Kenya, the Netherlands, Norway, United Kingdom—mentioned three groups.

Indigenous people were most often mentioned in reports of states from the Americas and Oceania while States Parties from Europe more often linked refugees, Roma, or asylum seekers with housing. However, 53 states (43%) did not mention any of the four groups in the housing context. These were especially the states from Africa and Asia.

6.1.4 The Committee's View

The Committee represents the second site of the discourse. In contrast to the ICESCR member states, the Committee was and still is not obliged to speak in a full scope about all ESC rights, including the right to housing, in each of its Concluding Observations. Instead, it has the opportunity (one could also say: the obligation) to raise the issues it regards as necessary. In sum, the number of inadequate housing issues in the Concluding Observations was significantly lower than in States Parties reports. The Committee mentioned homelessness in its Concluding Observations that were addressed to 89 of 130 states (68%). With respect to Spaces of Inadequate Housing, 78 states (60%) were addressed and 42 (32%) with respect to street children. Finally, 22 states (17%) were neither addressed about homelessness nor about Spaces of Inadequate Housing. Comparing both sides of the discourse with each other, the degree of attention afforded to the main forms of inadequate housing was considerably lower in the Concluding Observations than in the States Parties reports.

In terms of housing for disadvantaged and vulnerable groups, the situation is different and a little bit complicated. Refugees were the most frequently mentioned group in the States Parties reports, and Roma were regarded as a housing issue (see above); the Committee also talked about refugees most (53% of 130 addressed states), but when it talked about housing, it talked more often about housing for indigenous people: 29 states (22%) were addressed on this topic while 25 states (19%) were addressed with respect to housing for refugees. The Committee's attention to Roma (26 addressed states, 20%) was on a similar level, but a little bit lower for asylum seekers (19 addressed states, 15%). It seems that the Committee regarded indigenous people as a more vulnerable group with respect to housing than the other three groups.

In all, this statistical overview has shown that many states worldwide reported inadequate housing, albeit with different levels of attention and different focal points. States from the UN region of Asia reported on inadequate housing with considerably less attention than states from other regions. Street children were reported upon worldwide, particularly by states from Africa and the Americas.

This section has proven that most states acknowledged housing as an important part of ESC rights in their reports. However, the numbers above only show if and how often the states and the Committee talked about issues of inadequate housing. The next section discusses the states' focal points.

6.2 Comparison: Each State Is Unique, But...

Of course, each state is unique in the way of reporting ESC rights, including the human right to housing. Nonetheless, some states differ from others in certain aspects, for example, with respect to poverty, social citizenship, the focus on homelessness or Spaces of Inadequate Housing, or because they did not pay attention to housing at all.

I sorted the states into different groups (types) based on their way of talking about their implementation of the human right to housing. This is, from a methodological point of view, quite tricky for some reasons.

First, the data base for each ICESCR member state is different. Some states started to submit their reports in the 1970s while other states joined the ICESCR community many years later and have so far only submitted one report. Second, each state's view on housing could have changed during the past few decades. Third, many reports were and still are delayed, evasive, inconclusive, or lack data (U. Davy 2014: 243; see above). These aspects had to be kept in mind. Nonetheless, every state of the world community had the opportunity to ratify the ICESCR since 1966. Since 1976, when the Covenant came into force, every ICESCR member state had the opportunity to submit reports. In each of these reports, the states were and still are obligated to fully describe their implementation of ESC rights. Identifying the dominating view in a states' reporting is possible for states that submitted six reports as well as for states that have so far submitted only one report.

Sometimes, states were distinct because they showed not one, but two main qualities in their reports, such as when they primarily concentrated on one main form of inadequate housing and talked about their housing policy in a social citizenship language. So the reader might keep in mind that the boundaries between the different types blur, and most states could be put into one group, but some few states were put into two. I generated four groups: States with low attention, states that focused on homelessness or Spaces of Inadequate Housing, states that raised their poverty, and states that entered the road to social citizenship. Nevertheless, 34 states remained that I could not put into any of these groups. For them, I generated a fifth group: states with several or no focal point. Their reporting was interesting for different reasons and shall be briefly presented in the following section.

6.2.1 States with Several or No Focal Points

The group of the states 'with several or no focal points' encompasses all States Parties that reported inadequate housing with at least a medium level of attention, but did not fit into any of the other groups: Albania, Argentina, Belgium, Chile, China, Colombia, Cyprus, Czech Republic, El Salvador, Georgia, Guatemala, Iraq, Ireland, Israel, Jamaica, Malta, Mauritius,

Mexico, Morocco, New Zealand, Panama, Peru, Portugal, the Republic of Korea, Russia, Solomon Islands, Sweden, Thailand, Trinidad and Tobago, Tunisia, Uruguay, Venezuela, Vietnam, and Yemen.

The states from this group are distributed along all UN regions, but mainly Latin America and the Caribbean, Europe, and Asia. Of course, the states showed differences and overlaps in their reporting. States from Latin America and the Caribbean reported inadequate housing in considerable detail: Colombia (1994, 2000, and 2008), Mexico (1988, 1992, 1997, and 2004), Panama (1999), and Trinidad and Tobago (2000). Chile's quotation from 1979 summarizes the view of Latin American countries on responsibilities quite well:

> The State recognizes and endorses the fact that families acquire housing through their own efforts and savings.
>
> By this principle, housing is no longer a State gift, the fruit of the sacrifices of many used to benefit a privileged minority—a view which has condemned a large part of our population to live in unnecessarily poor conditions. Rather it is the fruit of individual efforts, supported by the State in the case of the lowest income sectors.
>
> [...] The State should play a secondary role in housing. It is for the private sector to marshal resources and means to meet aspirations for housing. (Chile 1979: 6)

Colombia (2008: para. 690 and 709) mentioned housing as a part of its constitution and reported housing for displaced people and land for indigenous people (Colombia 2000: para. 348). The state talked a lot about supranational actors involved in its housing policy, but not directly with respect to inadequate housing. Portugal is the only State Party that reported homeless people selling street newspapers:

> Among the support projects for the homeless, mention should be made of the CAIS (Support Group for Integration of the Homeless) project, begun in 1995. This is a community support association which publishes a monthly journal sold exclusively by homeless people or by people belonging to other socially-excluded groups. The vendors receive 80 per cent of the income from the sale of the magazine, which provides them with a

> respectable job and a small wage enabling them to eke out a meagre living. The founders of the project regard it not as an end in itself, but as a transitional measure that will encourage the vendors to take further steps towards reintegration into society. Among the project's backers are the Lisbon city council and the State Secretariat for Young People. (Portugal 1997: para. 418)

This report on homeless people selling newspapers serves as one of few examples in which homeless people were regarded as entrepreneurs who, by joining the (usually) informal working market, 'take... steps towards reintegration into society.'

A few states mentioned urban migration as the main cause for Spaces of Inadequate Housing (e.g., Portugal 1983). Others linked the incidence of homelessness with unemployment, alcoholism, or drug addiction (Israel 2009: para. 443). Russia (Russian Federation 1995: 208 and 251; 2001: para. 277–284) elaborated more on statistical information than on causes, policies, or measures. Some states were distinct because they cited territorial conflicts or natural disasters as central when they reported causes of inadequate housing. Cyprus reported the 'Turkish invasion' and the consequences of its housing policy in the beginning of the ICESCR discourse (Cyprus 1979), many years later (Cyprus 1989 and 1996), and in its most recent report, more than 30 years after the incident.

> The emphasis of the Government housing policy today continues to focus on population displaced following the 1974 Turkish invasion. But the Government recognizing the social and regional dimension of housing policy, has also introduced several Housing Programmes for married couples in specific areas, for large families, for public assistance receivers etc. All programmes lead to the acquisition on one housing unit per family or the repair/extension of the existing house. The form of assistance offered through these Programmes can be a long-term low-interest loans, grants, rent subsidy etc. (Cyprus 2007: para. 305)

Other States Parties whose housing reports were particularly impacted by wars and other armed conflicts are Iraq (1995) and Georgia (1997). Georgia reported inadequate housing caused by military conflicts and other incidents:

> The natural disasters (earthquakes, floods, avalanches), military conflicts, and the fuel/energy crisis that have struck the country since, the depreciation and decay of utility networks and structures through standing idle and lack of maintenance, together with a range of other subjective and objective circumstances, have adversely affected the quality of the housing stock and left it in still worse condition. (Georgia 1997: para.193)

El Salvador's manner of reporting was similar to other Latin American reports. The country used liberal language and stressed the importance of the family as 'the basic unit of society and as an important agent for social, political and cultural change' (El Salvador 1994: para. 167). Moreover, El Salvador mentioned land titling (El Salvador 1994: para. 203–204). The state's reporting was strongly influenced by naming the consequences of natural disasters such as hurricanes, droughts, or earthquakes (El Salvador 2004: para. 5). These incidences heavily influenced the country's housing policy (El Salvador 2004: para. 526–692). The country stressed adequate housing through land ownership for families:

> The basic purpose of the 'El Salvador, Land of Owners' programme is to provide legal certainty by awarding title on an individual basis and swiftly, efficiently and safely to the plots occupied by thousands of poor families in marginalized areas, 'pirate' settlements, unauthorized communities, etc., and entering the titles in the Public Land Register. Title has been awarded to 1,629 families living in 37 communities in eight of the country's departments. On the basis that an average family has five members, this programme has helped 8,145 Salvadorans to become authentic and legitimate owners. (El Salvador 2004: para. 663)

EL Salvador talked often about an involvement of supranational actors, which it also justified with the consequences of natural disasters (El Salvador 2004: para. 677–678 and 681).

States that reported natural disasters or territorial conflicts frequently talked about an involvement of supranational actors. Causes of inadequate housing also became a supranational issue in the case of territorial conflicts. When states reported natural disasters, supranational actors were often involved as donors. In the case of territorial conflicts, supranational actors—namely other states—were also mentioned in the states' discussions, although not as helpers, but rather as guilty parties.

New Zealand reported homelessness and Spaces of Inadequate Housing with considerable detail, but set itself apart because the country concentrated on Maori people in its housing sections (New Zealand 1990: 419–424; 2001: para. 361–365). Israel (1998: para 459–482; 2001: para. 342–376; 2009: para. 474–522) concentrated on emigrants from Ethiopia and on Bedouins living in illegal settlements. By doing so, the country conveyed their infrastructural measures, such as for the water supply (Israel 2009: para 520–522). As presented previously, Israel (2001: para. 359) is one of the countries that openly contradicted the Committee.

Morocco, when reporting about a program for a 'wider range of socially disadvantaged groups living in insecure circumstance,' divided between 10 different groups:

> [W]omen in highly insecure circumstances; homeless young people and children living in the streets; former prisoners without means; abandoned children; destitute older persons; homeless persons with mental disabilities; beggars and vagabonds; persons with HIV; drug addicts; and destitute persons with disabilities. (Morocco 2013: para. 22)

All in all, the section has once again shown that housing is a complex policy field meaning that there is a long list of different causes for inadequate housing, numerous different vulnerable and disadvantaged groups, and of course, measures to respond to inadequate housing.

6.2.2 States with Low Attention

I divide the states with low attention into two sub-groups, the nearly silent states and the silent states. Diffusion Map 6.1 (p. 169) shows some geographical concentrations, that is, large African states from the North of the continent, states from southern or eastern parts of Europe, and states from Central Asia and Eastern Asia.

More than one fourth of the 129 states, namely 35, reported on inadequate housing with little detail. These are the nearly silent states: Algeria, Austria, Belarus, Bolivia, Bulgaria, Chad,

Croatia, Democratic People's Republic of Korea, Egypt, Honduras, Iceland, Italy, Japan, Jordan, Kyrgyzstan, Latvia, Libya, Liechtenstein, Luxembourg, Madagascar, Mauritania, Republic of Moldova, Monaco, Mongolia, Montenegro, Nigeria, Serbia, Slovakia, Slovenia, Suriname, Switzerland, Syria, Tajikistan, Togo, and Uzbekistan.

These States Parties gave at least some attention to inadequate housing. Sometimes they only reacted to their reporting obligation with some basic statistical information (e.g., Switzerland 1996: para. 484). Luxembourg (1988: para. 58) reported housing as the most expensive item in the state's budget. Others only reported a lack of statistical data on homeless people, illegal residences, or evictions (e.g., Japan 1998: para. 166). Liechtenstein simply reported that '[h]omelessness does not exist in Liechtenstein. A facility for homeless persons had to be closed for lack of use' (Liechtenstein 2004: para. 167).

Six states from the group were distinct insofar as they only reported one main form with a low or medium level of attention (homelessness, for instance) and did not mention any other (Spaces of Inadequate Housing). Madagascar and Syria are such countries. Syria (Syrian Arab Republic1989: para. 76) reported that no one should be left homeless on Syrian territory. Ten years later, the country spent a few paragraphs on street children (Syrian Arab Republic 1999: para. 144–148). Madagascar mentioned street children in only a few paragraphs as well, often as one group among others: 'State assistance—reinforced by international associations—to poor families, street children, orphans and prisoners, among others' (Madagascar 2007: para. 488). In the reports of Croatia (2000), Honduras (1998 and 2014), Kyrgyzstan (2013), and Serbia (2011), it worked the other way around. These countries mentioned Spaces of Inadequate Housing, albeit not with very much attention, and did not talk about homelessness.

Nine states neither mentioned homelessness nor Spaces of Inadequate Housing with any word: Barbados, Gabon, Kazakhstan, Kuwait, Lebanon, Romania, San Marino, Senegal, and Turkmenistan. I call this sub-group the silent states. The group of the silent states would be larger if I had taken all the states into account that ratified the ICESCR, but have not yet submitted a report (e.g., Bangladesh, Papua New Guinea, or South Africa). The silent states can be again divided into three types.

The first type includes states that not only avoided reporting any form of inadequate housing, but also did not talk about housing at all or only in a few words. This applies to Barbados and Lebanon. I cannot satisfactorily explain why these countries did not mention any form of inadequate housing. Barbados submitted its last report in 1983 and Lebanon in 1993. These countries did not take their reporting obligations seriously. This is, however, neither an excuse nor a satisfactory explanation for not mentioning inadequate housing.

The second type includes states that recognized adequate housing as a state's duty and talked about policies, construction programs, plans, and other measures in the field of housing. This applies to Gabon, Kazakhstan, Kuwait, Romania, Senegal, and Turkmenistan. These states gave the reader the idea that they had some problems with inadequate housing that they were trying to solve, but they did not explicitly admit to any form of inadequate housing. Romania (2011: para. 263–272) did not mention inadequate housing as a subject, but reported in detail on its social housing policies for Roma communities and evicted persons.

San Marino comprises the third type. The state talked about housing, but absolutely did not give any sign that inadequate housing could be a problem for its population. In the Concluding Observations considering San Marino's States Parties reports, the Committee also seemed to believe that there was no problem with respect to housing in San Marino: 'The Committee notes with satisfaction that the right to adequate housing is respected in San Marino, and that 80 per cent of the population own the housing units in which they live' (CESCR_SMR 2008: para. 6).

6.2.3 States Focused on Homelessness or Spaces of Inadequate Housing

Spaces of Inadequate Housing and homelessness are the main codes to analyze inadequate housing in the States Parties reports. Of the 129 reporting States Parties, 107 mentioned homelessness in at least one of their reports; 104 States Parties mentioned Spaces of Inadequate Housing, and 91 states reported both main forms. However, some states were distinct because they reported only one of the two main forms of inadequate housing and neglected the other (this includes six of the 'nearly silent states'; see above), or they reported both forms, but paid considerably more attention to one over the other. In summary, more than one fourth of the reporting states either focused on homelessness or Spaces of Inadequate Housing in their reports.

Diffusion Maps 6.6 and 6.7 illustrate the global distribution of the states' main reported housing form. European states and Australia focused on homelessness. African and Asian states emphasized Spaces of Inadequate Housing. In Latin America and the Caribbean, Spaces of Inadequate Housing dominated as well.

Fifteen states paid more attention to homelessness than to Spaces of Inadequate Housing (including Madagascar and Syria from the group of the nearly silent states and Poland from the group of states on the road to social citizenship): Australia (1992 and 1998), Cambodia (2008), Denmark (1996, 2003, and 2010), Estonia (2001 and 2008), Finland (1995, 1999, 2005, and 2011), the Former Yugoslav Republic of Macedonia (2005), Germany (2000 and 2008), Hungary (2005), Islamic Republic of Iran (2009), Lithuania (2002 and 2010), Madagascar (2007), Norway (2004 and 2010), Poland (1996, 2001, and 2007), Syrian Arab Republic (1999), and Ukraine (1999, 2006, and 2011). Usually, these countries started to develop their focus in the second or third reporting period, while their early reports usually did not confirm such a focused orientation.

The states discussed different measures to respond to the needs of homeless persons. Some of them emphasized spatial solutions like soup kitchens or night shelters (e.g., Estonia 2001: 508 and 542) while others mentioned housing allowances (e.g., Germany 2008: para. 247) or social support measures (e.g., Australia 1998: para. 186–187). Hungary reported homelessness and facilities to claim other needs beyond housing:

> The development of the services of institutions currently described as day shelters, and integration of already existing models are of key importance, because only a fragment of homeless people use the accommodation facilities, thus the best opportunity for establishing contact could be provided in a day care system with a wide range of services. In addition to the bathing,

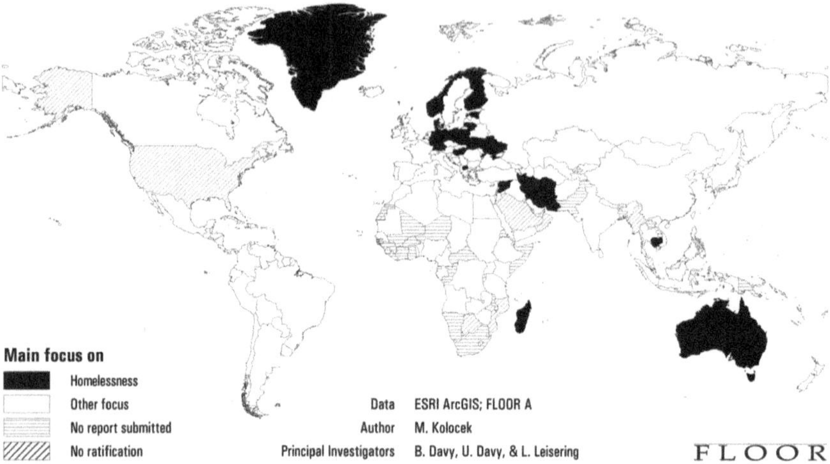

Diffusion Map 6.6 ICESCR member states that focused on homelessness

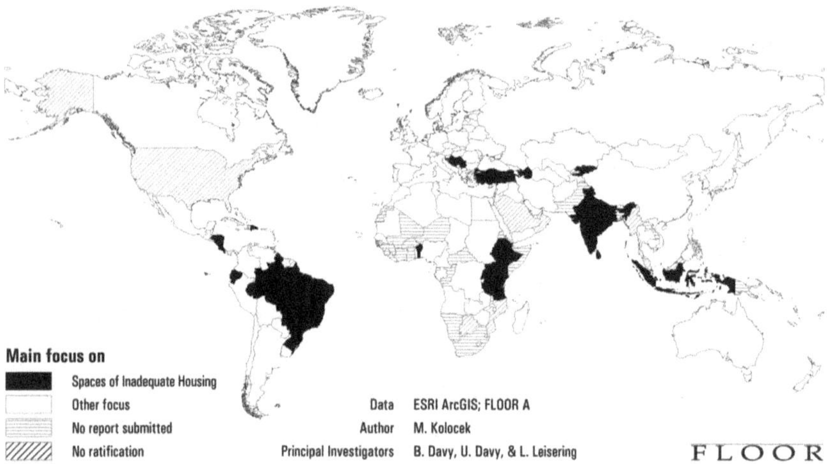

Diffusion Map 6.7 ICESCR member states that focused on Spaces of Inadequate Housing

washing and cooking opportunities involved in the currently available day services we would like to provide information services, luggage guarding services, to supply a postal address, legal assistance, access to computers and Internet, job search and training. (Hungary 2005: para. 408)

Hungary mentioned a day care system with a 'wide range of services.' The state presented a kind of realistic view by explaining that only a fragment of homeless people used accommodation facilities. The quotation serves as a nice example of how de-commodification of land use for homeless people could look like. Germany's reporting on homelessness demonstrated a few citizenship ingredients, but when it came to asylum seekers, the state's talk went in the opposite direction. In the late 1970s, the Federal Republic of Germany (1979: 20) reported that '[s]ocial assistance is not only paid to German nationals. Homeless foreigners as well as recognized refugees and persons entitled to asylum are treated like German citizens.' However, 21 years later, the state presented a new policy and a different view:

> The Federal Social Assistance Act is no longer applicable to asylum-seekers and other foreign refugees who do not have consolidated resident status in the Federal Republic of Germany. Such persons now receive benefits under the Act on Benefits for Asylum-Seekers, which came into force on 1 November 1993 and was amended on 26 May 1997, and again on 25 August 1998.
>
> [...] The Act on Benefits for Asylum-Seekers gives priority to benefits in kind so as to ensure that the benefits provided are actually used to meet the needs that the Act addresses. Granting benefits in kind instead of cash ensures that beneficiaries under the Act cannot use them to support people they have left behind in their countries of origin or to pay smugglers while themselves being left without sustenance. Furthermore, granting non-cash benefits makes Germany less attractive to refugees not subject to political, racial or religious persecution. (Germany 2000: para. 197–198)

In these quotations, Germany seemed to regard asylum seekers as potential rights violators, as people who wanted 'to pay smugglers,' a view that contradicts a modern understanding of global social citizenship.

Many different actors responded to homelessness, but market actors and supranational actors responded more to Spaces of Inadequate Housing. Of course, there were exceptions. Norway (2004: para. 270–273), for instance, stressed the division of roles between the states, municipalities, and the private sector. Some states reported almost solely on street children when they reported homelessness. The subject of street children proved interesting because states were more likely to report an involvement of supranational actors than when they talked about adult homeless persons (Table 6.4, p. 178). The Islamic Republic of the Iran (2009: para. 298.11) mentioned international assistance by the UNESCO in its street children policies.

Twenty-four States Parties paid considerably more attention to Spaces of Inadequate Housing than to homelessness (including four states from the group of the nearly silent states): Azerbaijan (1996 and 2003), Benin (2006), Bosnia and Herzegovina (2004 and 2010), Brazil (2001 and 2007), Costa Rica (2006), Croatia (2000), Djibouti (2010), Dominican Republic (2008), Ecuador (1989, 2002, and 2009), Ethiopia (2009), Guyana (1995 and 2012), Honduras (1998 and 2014), India (1983 and 2006), Indonesia (2012), Kenya (2006 and 2013), Kyrgyzstan (2013), Nepal (2006 and 2011), Nicaragua (2007), Rwanda (1985), Serbia (2011), Sri Lanka (1996 and 2008), Turkey (2008), Uganda (2012), and United Republic of Tanzania (1979 and 2009). Brazil is the only country from this group that also entered the road to social citizenship.

Some states from this group also mentioned territorial conflicts or natural disasters as causes for Spaces of Inadequate housing. Azerbaijan (1996 and 2003) reported 'the Armenian's aggression.' In its second periodic report, Sri Lanka's (2008: para. 42–61) reporting focused on displaced persons caused by a tsunami in 2004 and the consequences of an armed conflict with a terrorist organization. Similar to the formerly discussed sub-group, the states developed their focus mainly in the second and third period when the attention to housing increased. In the early 1980s, India (1983: para. 123) reported that the right to adequate housing was not provided for as a constitutional or statuary right, but was an accepted objective in terms of the national plan for social and economic development. In the third period, however, India (2006: para. 406) reported the right to shelter recognized as an integral part of the fundamental right to life under its Constitution. Nepal's discussion of housing focused on (political) refugees:

> Nepal has been providing shelter to the refugees since the beginning of time to those fleeing from their country for noble causes on humanitarian

> ground. Many people from neighbouring countries e.g., India, Tibet/China, Bangladesh, Bhutan, Burma and others have taken refuge in Nepal from time immemorial. Among them, the issue of Bhutanese refugees is paramount followed by Tibetan refugees. (Nepal 2006: para. 68).

States that focused on Spaces of Inadequate Housing also differed in terms of the measures and policies they presented. Most of them reported spatial solutions, for instance, slum upgrading projects (e.g., Kenya 2006: 119) or other infrastructural measures (e.g., Croatia 2000: para 338–339; United Republic of Tanzania 2009: para. 101–108; Turkey 2008: para. 423–424). Costa Rica (2006: para. 777–808) and Nicaragua (2007: para. 782–786, 808–810) stressed land titling as a solution. Prevention or social support measures only seldom came up. In contrast to homelessness, Spaces of Inadequate Housing were considered an issue for market actors and actors on the supranational level:

> The Kenya Informal Settlement Improvement Programme (KISIP) 2011–2016 has been introduced through a partnership between the Kenya Government and the World Bank to undertake tenure regularisation and installation of social and physical infrastructure in selected informal settlements and planning for urban growth in 15 municipalities. (Kenya 2013: para. 161)

Kenya reports about its cooperation with the World Bank. In the following paragraphs, the country describes implementing new social housing units and the provision of physical and social infrastructures in different parts of the country (Kenya 2013: para 162–166). The state, however, acknowledged that

> Kenya's housing market, particularly in urban areas, remains among the most expensive in Africa. Private developers have focused on housing for upper middle income and high-income groups. There is an under investment in low and middle-cost housing by both the public and private sectors. Demand for new housing units in urban areas currently stands at 200,000 units annually but only 23 per cent of this need is being met. This shortfall is giving rise to the proliferation of squatter and informal settlements and overcrowding. (Kenya 2013: para. 167)

The quotations show a typical dilemma that spatial planners and other housing policymakers must respond to. Improving the existing urban housing conditions (including the social and physical infrastructure) increases the demand for new, often low-income housing units that the private housing sector does not want to respond to and the public sector cannot cope with. The demand is often caused by rural-urban migration and results in overcrowding or new or growing informal settlements in urban areas.

As seen, in the case of Spaces of Inadequate Housing, tenure regularization and installation of social and physical infrastructure may combat inadequate housing in certain areas, but increase inadequate housing in others. In the case of homelessness, it could be similar. The states' night shelter policies were often regarded as appropriate solutions for homeless people. Nevertheless, they are at risk to be just another form of inadequate housing. Considering China's reaction to the Concluding Observations of the Committee, in which the country emphasized that people live in bed-space apartments out of choice (People's Republic of China 2003: para. 592), one could conclude: As long as there is free migration (rural-urban migration as well as transnational migration), the full implementation of the human right to housing, meaning that all people all over the world will live in adequate housing, is more than unlikely—it seems impossible. This is true for Spaces of Inadequate Housing and for homelessness. Some states have already started to deal with that. Apparently, no one can fulfill the human right to housing entirely without violating other human rights, the right to free movement for instance. Housing's interrelationship with other rights and needs also becomes visible when regarding the next group, namely states reporting their poverty in detail.

6.2.4 States Raising Their Poverty

U. Davy (2014: 259) discovered that the incidence of the term poverty generally increased over the years of reporting. My research findings confirm these results for the case of housing. Fourteen states concentrated on poverty in many parts of their reports, including the housing sections: Afghanistan (2007), Angola (2008 and 2014), Armenia (1997 and 2011),

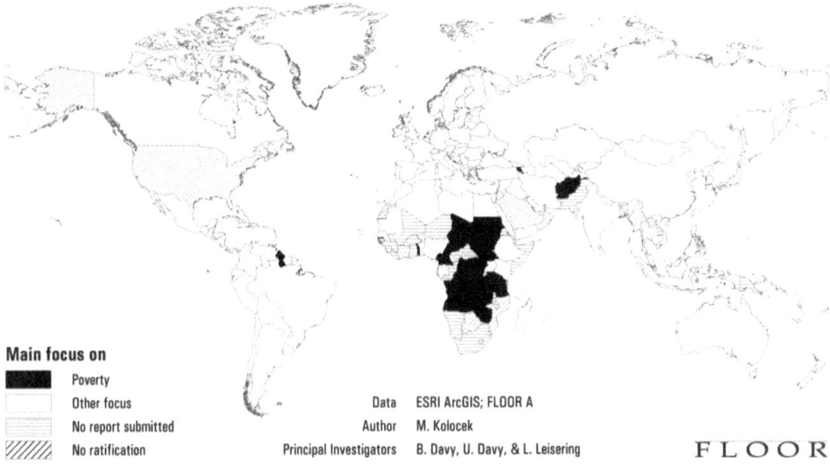

Diffusion Map 6.8 ICESCR member states stressing their poverty

Burundi (2013), Cameroon (2010), Chad (2007), Democratic Republic of the Congo (2007), Gambia (2012), Guyana (1995 and 2012), Sudan (2012), Togo (2010) United Republic of Tanzania (2009), Zambia (2003), and Zimbabwe (1995) (Diffusion Map 6.8).

Most likely, these states elaborated on Spaces of Inadequate Housing or on street children. Afghanistan, Armenia, and Guyana are the only non-African countries from this group. Afghanistan's (2007) short report dealt with the consequences of two decades of war.

> After two decades of war, the standard of living in Afghanistan is under chronic and severe conditions both in urban and rural areas. [...] Afghanistan is expected to be urbanised in the near future. In urban areas, the MUDH provides housing, services and employment in line with the Urban Reconstruction Plan, prepared with support of UN-Habitat. Donors including the World Bank, EU, Germany, Japan, the International Committee of the Red Cross, and Care are implementing housing supply, sanitation, and drainage projects. (Afghanistan 2007: para 93)

The state talked about expected urbanization and an involvement of many supranational actors that supported its policies with donations.

Housing supply and infrastructural measures were in the center of the reporting. In the third period, other states discussed social support, social integration, land titling, or other usual measures of responding to individuals' housing problems. However, war seemed to render a state's manner of reporting reminiscent of other states during the first period. Housing supply policies and infrastructural measures dominated. In contrast to the first period, though, in the case of Afghanistan, supranational actors played a significant role.

Armenia reported in the first State Party report about an earthquake in 1988 that destroyed one-sixth of the country's housing stock and left many people homeless (Armenia 1997: para. 208) or living in other inadequate housing conditions (Armenia 1997: para. 178–179). Moreover, the state was prominent in its preferences for privatization of housing, supported by the World Bank (Armenia 1997: para. 206–218). In the second report, the focus was more on sustainable development (Armenia 2011: para. 250) and housing for returning refugees exiled from Azerbaijan (Armenia 2011: para. 270–274).

Member states that stressed their poverty often explained how poor they were in the sections concerning Article 11 ICESCR (e.g., the Gambia 2012: para. 117–119; Guyana 1995: para. 67; Zambia 2003: para. 186) or in the first paragraphs of their reports (e.g., Cameroon 2010: para. 47; Chad 2007: para. 36). They gave statistical information to what extent an economic crisis influenced their policies to guarantee an adequate standard of living (Democratic Republic of the Congo 2007: para. 188–196). Sudan (2012: para. 41–55) explained in detail the current situation of refugees in the country. The state reported that about 150,000 refugees live outside the refugee camps and do not receive any assistance from the international community, but rely on, as the state said, 'sharing with citizens the already minimal social services available in the area' (Sudan 2012: para. 43). Togo talked about poverty in different contexts when criticizing the consequences of structural adjustment programs (Togo 2010: para. 269–272 and 532), but paid little attention to inadequate housing.

Most states talked about an involvement of supranational actors who helped them to implement ESC rights.

> [R]ecognizing that extreme poverty still affects 18% of the population, the GoG [Government of Guyana] introduced specific programmes which target the very poor. [...] Through partnership with Habitat for Humanity and Food for the Poor, GoG reserves a specific number of houselots in the government housing schemes for the very poor who would not be able to pay the subsidized cost of a houselot nor build. These organizations then build standard houses for these families and arrange repayment at very concessionary rates with which they are able to comply. (Guyana 2012: para. 273)

In the following paragraphs, Guyana reported that the poverty rate is even higher in the rural areas and that gender differences are not playing a role in the distribution of poverty (Guyana 2012: para. 298 and 316). Supranational actors played an important role by cooperating with governmental actors, but also sometimes as donors. The states, however, did not conceal that the involvement may not always yield positive effects, particularly in the case of structural adjustment programs offered by the IMF or the World Bank (Cameroon 2010: para 3). Zimbabwe reported that

> five years ago the Government introduced the Economic Structural Adjustment Programme (ESAP) in terms of which it opened up the economy with a view to encouraging economic growth and development and thereby improving the quality of life of all the people. Recently, the Government introduced the Poverty Alleviation Action Plan (PAAP), which will supplement and cushion the effects of ESAP and which is intended to alleviate poverty and improve the quality of life. Both programmes are being undertaken with the assistance of donor countries (such as members of the European Union), IMF, the World Bank, etc. (Zimbabwe 1995: para. 9)

In this excerpt, Zimbabwe mentioned the Economic Structural Adjustment Program (ESAP), which has been heavily criticized by many experts. The country did not directly comment on the program's negative effects. Instead, the country reported quite neutrally the 'Poverty Alleviation Action Plan,' whose aim was to 'supplement and cushion the effects of ESAP.' Later in the report, the State Party acknowledged negative consequences:

> The Economic Structural Adjustment Programme (ESAP) has adversely affected the poor by eroding the gains in social services, particularly education for their children and health care provision, as a result of cuts in public expenditure and retrenchment of mostly the unskilled and semi-skilled workers.
>
> The stringent cost-recovery exercise introduced as a result of the ESAP puts the onus on the poor to prove that they are unable to pay and do not earn the stipulated Z$ 400 per month, leading many people to avoid use of the health services.
>
> Zimbabwe adopted an economic reform programme (Economic Structural Adjustment Programme or ESAP) which deregulated the economy by allowing a greater freedom of open market forces as well as cutting down on government expenditure on social services. As a result, prices of basic commodities and services were pushed up, and families' real income was generally eroded. (Zimbabwe 1995: para. 166, 167, and 207)

The problems with SAPs and similar programs were that they aimed for economic growth with the hope to increase the standard of living for almost all people, often through liberalization. With respect to social rights such as health care and education for the poor and other vulnerable and marginalized groups, this policy often failed. It increased inequality. The states recognized and acknowledged these negative effects, but only seldom complained about their international partners (and donors). As discussed, the Committee was also critical with SAPs (Section 5.3). However, despite the bad experiences, Zimbabwe and other states with structural adjustment programs (e.g., Guyana 1995: para. 3 and 70) did not indicate in their reports that they wanted to abstain from involving supranational actors in their housing policies. What changed were the policies and programs—the involved actors remained. Regarding the group of the states that were distinct because they raised their poverty, one could conclude that the globalization of social policy in these (albeit only few) reporting countries might have raised inequality.

6.2.5 States on the Road to Social Citizenship

Chapter 5 has demonstrated that, during the past almost four decades, ingredients—or, in the terminology for the present research: meaning patterns—of social citizenship have spread into the reporting of the ICESCR member states: Homeless people and Spaces of Inadequate Housing dwellers were more likely regarded as right holders than at the outset of the reporting, and housing measures became more targeted. The states began to link refugees, indigenous people, asylum seekers, Roma, and other vulnerable and disadvantaged groups with the right to housing, although on different levels. Since 2000, land rights for women are on the rise as well. Housing structures that had previously been regarded as illegal were increasingly termed informal. The states started to stress the economic potentials of informality by using land titling or microcredits. Equality goals spread into the housing policy talk of several ICESCR member states and were strongly demanded by the Committee.

Moreover, ingredients of a globalization of social policy appeared in the housing sections of the States Parties reports. More and more, the states mentioned an involvement of supranational actors when reporting their policies for adequate housing. Additionally, the states reported their participation in international conferences and their membership in international networks, their ratifications of international conventions and covenants. Non-national refugees and asylum seekers were also, albeit on a small level, increasingly linked with housing policies. Social citizenship became progressively more global.

Almost all States Parties that reported inadequate housing with more than a few paragraphs showed at least a few traces of social citizenship. Some States Parties showed, in comparison to other states, significantly more ingredients of social citizenship in the sections concerning the human right to housing than others. Of course, no state presented the ingredients of social citizenship in every paragraph of every submitted report. However, the differentiations for the now presented group can be described as follows: Several States Parties showed ingredients of social citizenship in their reports, but some showed considerably more ingredients than others. They shall be called states on the road to social citizenship. I also point out those states whose discussion additionally showed elements of a globalization of social policy. These states, one could say, entered the road to global social citizenship.

States on the road to social citizenship frequently paid plenty of attention to homelessness and Spaces of Inadequate Housing. They usually incorporated vulnerable and disadvantaged groups such as refugees, indigenous people, Roma, or asylum seekers into their reports on housing. These states employed ample measures, policies, and programs to deal with inadequate housing, and by doing so, confirmed their recognition of both the heterogeneity and the complexity of homelessness and Spaces of Inadequate Housing. By talking about their policies, programs, and measures, they regarded affected people as individuals—as right holders or as entrepreneurs, or both—and not as an anonymous mass. The measures, too, focused more on individuals, such as when the States Parties mentioned microcredits, land titling, targeted financial assistance, targeted health care, or integration policies. Often, the policies were discussed under the umbrella of equality as a policy goal. States on the road to social citizenship showed some kind of self-critical view when they acknowledged how difficult it was to reach equality in the housing sector. The Committee was nevertheless not afraid to critically comment on these States Parties reports in the Concluding Observations.

The group of states on the road to social citizenship consists of 11 states from each UN region except of Oceania and Africa: Brazil (2001 and 2007), Canada (1991, 1997, 2004, 2005, and 2012) Finland (1995, 2005, and 2011), France (2000, 2007, and 2013), Greece (2002 and 2012), the Netherlands (1996 and 2005), Paraguay (1998, 2006, and 2011), the Philippines (1994 and 2006), Poland (2001 and 2007), Spain (2002 and 2009), and the United Kingdom (2001 and 2007) (Diffusion Map 6.9).

Canada (1991: para. 112) is the only state reporting that the International Year of Shelter for the Homeless in 1987 (UN 1981) influenced its view on housing. The report recognized that homeless peoples' needs are closely associated with other social and health needs and that the affected require more than just shelter assistance (Canada 1991: para 113). The state paid a lot of attention to both main forms of

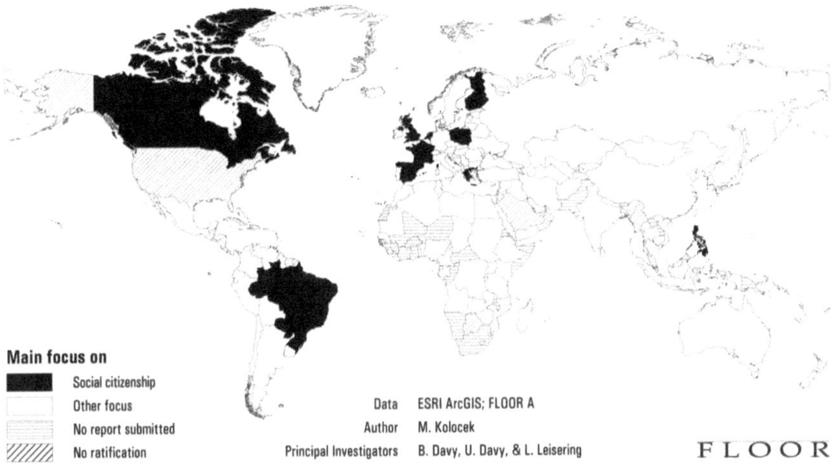

Diffusion Map 6.9 ICESCR member states on the road to social citizenship

inadequate housing, but regarded homelessness from different points of view. Canada emphasized integration goals, for example, by reporting how it pointed out the independence of homeless people (Canada 1997: para. 624) and called homeless people 'individuals' (Canada 2004: para. 341) and 'citizens' (Canada 1997: para. 1557). The country paid attention to housing for Aboriginal people (e.g., Canada 1991: para. 104) and refugees (Canada 2005: para. 84), and presented itself as a supranational actor when it talked about how the Canadian International Development Agency (CIDA) focused on street children in an international context (Canada 2004: para. 272). In the Annex of the country's recent report, Canada (2012) reported on a court decision of The British Columbia Court of Appeal. The Court held that 'tent city,' which was set up by 70 homeless people in a public park of Victoria, was not against the law. Instead, the Court 'found that the impugned bylaws and operational policy, which prohibited persons from erecting temporary shelters on public property, constituted an interference with the life, liberty and security of homeless persons' (Canada 2012: p. 114). The mention of the case in the report is remarkable and so is the decision as such. The British Columbia Court of Appeal had to accept a new form of informal housing because if it did not, other (human) rights such as the right to life, liberty and security would have been violated. Not respecting informal housing, so held the Court, would be against the law, in other words: illegal.

Most of the European states from the group of the states on the road to social citizenship, namely **Finland, Greece, the Netherlands, Spain,** and the **United Kingdom** showed not only a national social citizenship view on housing, but also elements of global social citizenship.

Based on the state's reporting, everyone in **the Netherlands** should enjoy the same rights and was entitled to protection and health care, education, and other essential services, and, as the State Party explicitly emphasized, this also included homeless people (the Netherlands 1996: para. 134). In the same paragraph, the Netherlands explained how it tried to guide young homeless people back to a roof over their heads, education, and employment and mentioned that refugees 'who have formal status are treated in the same way as any other ordinary Dutch citizen on the housing market' (the Netherlands 1996: para. 134). The key subject of inadequate housing in the reports of the Netherlands concerns squatters and illegal occupancy. While in its early years, the state talked about penalties to squatters (the Netherlands 1983: para. 83–85), in its last report from 2005, the country avoided the term squatters, but talked about illegal occupancy instead (the Netherlands 2005: para. 347). The **United Kingdom** reported both main forms in detail and presented its aims to enable homeless people to make their own decisions, live their own lives and become active members of the community (United Kingdom 2001: para 11.76 and 11.143). The state showed a self-critical view when it acknowledged that too many people dwelled in temporary accommodations (United Kingdom 2007: para. 150). Moreover, the state reported its housing policies for asylum seekers and refugees (United Kingdom 2001: para 11.17 and 11.88) and about funding UN-Habitat and the Cities Alliance (United Kingdom 2007: para. 68). **Spain** (2002: para. 68) mentioned being a member of a cooperation program that supported training and education for women from urban slums and from rural areas in Latin America. Moreover, Spain mentioned shelter for homeless people that provided 'guidance, assistance and the means necessary to restore their ability to find a place in society' (Spain 2002: para. 393). Spain emphasized its aim to guarantee equal access to housing for its citizens and its aim to secure

> for all families and citizens the freedom to choose the model of access to housing that best suits their circumstances, preferences, needs and economic situation, establishing that renting should be possible for the same income levels as those determined for access to property. (Spain 2009: para. 562)

Considering 'Gypsy communities,' the state reported the development of comprehensive welfare programs (Spain 2002: para. 168).

Finland reported on both main forms of inadequate housing carefully, but talked more often about homelessness. The state tried to integrate homeless people into the rental market (Finland 2005: para. 41) and mentioned a regional cooperation network supporting those homeless persons for whom access to housing was most difficult; the network tried to lead them back to independent living (Finland 2005: para. 45). The country's discussion was often marked by the aspiration to reach housing equality, for instance, when Finland reported that homeless persons receive the same health and medical services as other residents (Finland 1995: para. 311), or when Finland brought up its aspirations to integrate Roma people into the housing market:

> The Finnish Roma live in the same areas and in the same types of apartments as the majority population, and there are very few homeless Roma. However, the Roma are dependent on public rental housing for the reason that it is more difficult for them to find an apartment in the private rental market because of their weaker financial position and the prevailing prejudices. Although the supply of rental apartments has improved in the past few years, in respect of both public and private housing sectors, the Roma still face problems in the housing market. In the private housing market, the usually required guarantee deposits and the solvency requirement often make it difficult for Roma to find an apartment. In respect of private housing, it is rather difficult to intervene in possible discrimination because of problems of evidence. In respect of apartments granted on social grounds, the Roma are in an equal position with other applicants. The new Non-Discrimination Act (see information given under Article 2) further improves the possibilities of intervening in possible discrimination.

> The Ministry of the Environment, which is the authority responsible for housing, has drawn particular attention to the requirement of equality in its handbook for authorities deciding on the selection of tenants of rental apartments supported financially by the state. In cooperation with the Advisory Board for Roma Affairs, the Ministry of the Environment has also produced a handbook called 'The Special Aspects of Housing in Roma Culture'. It is meant for the local housing authorities and other relevant bodies, providing information on the Roma and thereby facilitating the access of the Roma to housing and their possibilities to change apartments. Both handbooks have been widely disseminated. Questions of equal treatment have also been addressed in various training and other sessions of local authorities and other bodies responsible for housing. (Finland 2005: para. 80–81)

The excerpts demonstrate the State Party's efforts to reach equality. Finland admitted problems for Roma people to find an apartment in the private rental market due to prejudices. The excerpts show that Finland regarded the private rental market more as a burden than as an opportunity, and the state told the Committee (probably incidentally) that it did not regard homeownership as the appropriate solution for Roma people. Instead, the state suggested renting. In a later paragraph, Finland (2005: para. 486) did not just explain the problems, but also demonstrated some self-confidence in mentioning that, in comparison to other States Parties, the housing integration of Roma had succeeded, although some problems remained. In its most recent report from 2011, the state replied directly to the Concluding Observations by the Committee (CESCR_FIN 2007: para. 11) and reported ownership and use of land in the Sámi Homeland (Finland 2011: para. 3–8). By considering its emergency shelters, the country acknowledged that these shelters do not meet the minimum standard, and therefore, should be replaced by other permanent housing solutions:

> Finnish homelessness policy is based on the notion that emergency shelters and residential homes, intended as temporary accommodation for the homeless, do not meet the minimum standard that should be expected of housing solutions for homeless people, and should therefore

> be replaced with better-standard housing solutions. The emergency shelters that still exist are gradually being converted into reception and assessment units for the homeless. Their key task is to find a more permanent housing solution for each homeless person and organize the necessary care and support services under the 'Housing First' principle. (Finland 2011: para 246)

There is no doubt that Finland's reporting shows many ingredients of global social citizenship. The Committee was nevertheless quite critical when responding to the States Parties reports. In its recent Concluding Observations, the Committee criticized the continuing shortage of social housing and the state's realization of the right to housing for Somali and Roma (CESCR_FIN 2014: para. 24), just to mention one example.

Greece (2002: para. 400–407) paid a lot of heed to the consequences of natural disasters. The country talked about housing for refugees (Greece 2002: para. 341 and 361) and asylum seekers (Greece 2012: para. 183 and 188) and described a program for housing for the Roma in detail (Greece 2002: para. 386–39 and 2012: para. 191–209). The state was distinct due to its special integration policy of illegal housing (Greece 2002: para. 323), but also because Greece discussed its view on the relationship between homeownership and renting in detail:

> Greek housing policy clearly favours the owner-occupier sector and the efforts of households to move into this sector away from rent. In this respect, renters can be considered a relatively disadvantaged group. However, in contrast to most European countries, the status of renters has not historically been associated with any particular social class. Mostly, it is associated with the stage in the life-cycle of the household. Given that mortgage financing has played a comparatively limited role throughout the post-war period, a lot of younger households without substantial parental assistance had to save for extended periods before moving into owner-occupation. (Greece 2002: para. 345)

Greece regarded renters as a disadvantaged group, although in relative terms. The country pointed out that 'the status of renters has not historically been associated with any particular social class.' The state stressed the life-cycle of households to explain why younger households did not live in owner-occupied houses. Also of high interest is the country's interpretation of its own Constitution:

> The Greek Constitution states in article 21, paragraph 4, that: the acquisition of housing by those that lack it or are inadequately housed, is the object of special attention by the State.
>
> This general statement does not, obviously, institute a full right to adequate housing in the strict legal sense of the term. On the other hand, it can be argued that it offers a base for legitimate claims by those in need, though such claims may be disputable as to the extent and form of the 'special care' the State is obliged to provide. Unfortunately, there are no specific laws or institutions that would further specify and implement this constitutional article in a legally rigorous way.
>
> Moreover, this article leaves open a number of controversial issues as to the nature of public care for housing. First, the term 'acquisition' predisposes housing policy in favour of owner-occupation and, therefore, may discriminate against support of other tenure forms. Second, the rest of the paragraphs in article 21 concern public care for the family, motherhood, childhood and the young, general health and the care of the invalid, the old and the poor. Thus, the housing paragraph is located amongst directives with a strong emphasis on social welfare and those most in need. In fact, the 'Popular Housing' programme in the Ministry of Health and Social Welfare, introduced in the 1950s and active up to the early 1980s, operated essentially in this spirit: it catered to the very poor, the homeless, refugees, and victims of natural disasters. It is interesting to note, as a sort of legal precedent, that the legal and administrative framework for the programme stated quite clearly who was a beneficiary, therefore establishing a right, but it also stated that provision was dependent on the resources available as determined by medium-term planning and fiscal restrictions. This philosophy appears to be prevalent in the Greek housing policy context, with a case in point the OEK programmes.
>
> Thus, the Greek Constitution and legal/political tradition form a rather flexible context that, on the one hand, obliges the State to assist Greek households in their efforts to acquire a decent house and, on the other, provides a political basis for demands by those in special housing need. This does not constitute clear-cut enforceable housing rights. As a result, there is a clear institutional gap in the Greek housing system in this respect

> even for individual cases of extreme need or homelessness: there are no legal avenues for demanding some form of shelter within a set time period or protection against eviction when rent payments cannot be met (though often judges substitute for social policy and give lenient rulings in case of extreme need). (Greece 2002: para. 347–350)

To summarize the quoted excerpt: Those who lacked the acquisition of housing, or were inadequately housed, enjoyed special attention from the state, but this does not mean that the state regarded adequate housing as a constitutional right. Instead, Greece acknowledged that the article 'leaves open a number of controversial issues as to the nature of public care for housing.' This applies to the right understanding of the term acquisition as well the answer to the question: Who is 'most in need'? Greece acknowledged 'a clear institutional gap in the Greek housing system,' and by doing this, showed how difficult equality is to reach in the case of housing. However, Greece made many efforts to fill this gap and interpreted the flexibilities in the disadvantaged groups' sense. In the case of homelessness, Greece reported many actors involved:

> A committee has been formed, composed by the National Social Solidarity Centre (EKKA) and other competent local authorities and stakeholders (municipalities, church, NGOs etc.), which has been entrusted with the planning and implementation of a country-wide data collection on the homeless, on the basis of the relevant European methodologies. It has also been given the task to elaborate a package of measures for the support of the homeless. (Greece 2012: para. 147)

The state reported about the European influence by mentioning 'European methodologies.'

France and Poland regarded inadequate housing from a more national point of view. Nevertheless, the countries' talking showed numerous ingredients of social citizenship. France repeatedly stressed its goal to combat the social exclusion of homeless people (e.g., France 2007: para. 83–87 and para. 312) and talked a lot about inequalities in general (e.g., France 2007: para. 88–93). The country reported thoroughly on the causes of homelessness (France 2007: para. 193–199) as well as different measures that had been taken (France 2000: para. 535–359 and 2007: para. 303–317). France talked about 24-hour reception and advice centers and a free telephone advice line (France 2000: para. 538). Moreover, the state reported having long-term measures for integration (France 2007: para. 307). France regarded homeless people as right holders, who, according to the report, just need to know their human rights:

> Since 2005, the approach to aiding the homeless has shifted towards the aim of achieving lasting reintegration of the individual, through suitable long-term accommodation, vocational training, regular monitoring of progress and information on fundamental human rights. (France 2007: para. 327)

France also reported self-critically about an incident in the winter:

> when a group of people decided to call on Parisians to come and share the tents of the homeless as a gesture of solidarity and protest against the inadequate response to the problem by the authorities. This mirror can also present a distorted image by tending to focus attention on housing problems whereas the issues involved are much more complex. (France 2007: para. 130)

Moreover, homelessness was given considerable attention by the media (France 2007: para. 304). The State Party acknowledged that emergency shelters could, at times, lack adequate housing, and therefore, France (2007: para. 301–311) tried to reach permanent housing solutions and other long-term integration measures. Like Finland, the state reported 'not to continue to increase the number of shelter places but to make every possible effort to help them to find real housing' (France 2013: para. 417). As said, the state's self-description of the policies designed to remedy inadequate housing showed a more national than an international (or global) perspective. The state did not turn its attention to housing for refugees and only briefly touched upon housing for asylum seekers. France neither mentioned any involvement of supranational actors in its housing policy nor did the state itself report acting as an international actor. Instead, by talking about its new integrated housing policy 'Housing First,' France (2013: para. 418) acknowledged that the state has chosen to continue implementing it under national stewardship while the European partners Finland, Denmark, and the United Kingdom opted for decentralization.

Poland's citizenship talk showed some elements of de-commodification of land use for homeless people. Poland, as many of the previously discussed European states, reported integration and equality goals in its housing policy (Poland 2007: para. 31 and 580) and described how it tried to prevent the exclusion of homeless people (Poland 2001: para. 191) and non-nationals (Poland 2001: para. 365). Poland (2007: para 575–576) mentioned funds allocated to NGOs that should respond to homelessness:

> The organizations run night shelters, shelters, homes for single mothers, food banks, treatment and nursing establishments, hospitals and hospices, homes for refugees, vocational workshops, hostels for violence victims, communities, re-adaptation and protected housing establishments, Mediation Centres. Emergency assistance centres have been also organized, and they include: warming shelters, eating-houses, kitchens for the poor, points of medical and sanitary aid, points offering assistance in kind, daily shelters, clubs, legal, psychological and family assistance points. (Poland 2007: para. 577)

Although the state did not explicitly stress this, one could assume that the services were offered for free. The scope of the assistance offered was reminiscent of the de-commodification of land use for homeless people that the students discovered by interviewing homeless people living in Hamburg (Ajayi et al. 2011).

Paraguay and Brazil are the two Latin American states that entered the road to social citizenship. **Brazil** (2001: para. 527) emphasized that it tried to reduce social inequalities by targeting its social policy measures toward the low-income population. The country showed a remarkable change of view with respect to informal and illegal housing (Brazil 2001: para. 509) and highlighted the right to the city (Brazil 2001: 352).

Additionally, the state talked about an involvement of supranational actors in its housing policies and mentioned, among others, UN-Habitat (Brazil 2007: 366), the World Bank, and the Inter-American Development Bank (Brazil 2001: para. 553 and 2007: para. 392). **Paraguay** (1998) first reported social citizenship in a more national manner, but then, in the following reports (Paraguay 2006 and 2011) also mentioned the involvement of some few supranational actors. The state talked about policies for Spaces of Inadequate Housing and homelessness and gave a great deal of thought to street children. The country also reported the legalization of land for indigenous people (Paraguay 1998: para. 23–26 and 2011: para. 5–16) and a program for social integration of indigenous street families (Paraguay 2006: para. 127). Additionally, the state mentioned housing loans for homeless people (Paraguay 2006: para. 427) and micro-enterprises for families living in precarious, unhealthy, and overcrowded conditions (Paraguay 2006: para. 291–297). In 2011, the state reported a social program for families living and working on the street which was supported by United Nations International Children's Emergency Fund (UNICEF) (Paraguay 2011: para. 107). Moreover, the state reported a pilot project of mutual-aid housing cooperatives that, inter alia, enabled persons in extreme poverty to obtain complementary incomes helping them to meet their basic needs and to improve their human capital in terms of education, health, and work skills (Paraguay 2006: para. 298–304). Housing demand for people living in poverty and extreme poverty was one of the main challenges in 2011 when the state reported its flagship housing improvement program:

> The objective of the flagship housing improvement programme is to reduce the housing deficit faced by families in poverty and extreme poverty in urban and rural areas, as well as indigenous communities, through the construction and improvement of housing, expansion of basic services and regularization of land tenure. (Paraguay 2011: para. 191)

Later in the text, Paraguay (2011: para. 193) noted that the state has invested little to solve the country's housing problem. Such a self-critical comprehension could be read in different ways. The state did not explicitly say that market actors (or other actors) caused housing problems. The state did not mention rural-urban migration either, but avoided reporting causes for inadequate housing in its last and second to last report. In the first State Party report, however, the state openly reported that the housing conditions in the rural sector were worse than in the urban sector which caused internal migration and an increasing demand for urban housing (Paraguay 1994: para. 297–298).

In the group of states on the road to social citizenship, the **Philippines** is the only country from Asia. The Philippines stood out due to its emphasis on private property as a cornerstone of its social policy. Moreover, the country is one of the few that talked about the World Bank's involvement in its titling policy (the Philippines 1994: para. 391). From the beginning of the State Party's reporting, security of tenure was an important element (the Philippines 1977: 2) that was later also mentioned in the country's Constitution (the Philippines 1983: para. 108). The state continuously talked in a liberal manner:

> In the promotion of social justice the State shall create economic opportunities based on freedom of initiative and self-reliance. The State shall undertake, in cooperation with the private sector, a continuing programme of urban land reform and housing. The State shall make available at affordable cost decent housing and basic services to underprivileged and homeless citizens in urban centres and resettlement areas. In the implementation of such programme, the State is required to respect the rights of the small property owners. [...]
>
> The Philippine Constitution protects the right to property and the right of the Filipino people to expect the Government to promote urban land reform and housing. In its effort to improve the plight of the urban poor, the Government must reconcile the seemingly conflicting rights of squatters and property owners. The rights upheld by the Constitution are limitations on Government to prevent arbitrary use of its power. Harmonization of the laws and policies of Government is therefore an imperative.
>
> [...] The Philippine Government encounters numerous difficulties in reconciling the rights of property owners, the right of the people to

> adequate housing and the need for humane treatment of those affected by evictions with its obligation to promote economic development by providing basic services and infrastructure. The process of fulfilling such obligation sometimes impinges upon private rights which the Government is also obliged to protect. (The Philippines 1994: para. 330, 362, and 365)

In this excerpt, the state and the market were regarded as the key responsible actors in the promotion of social justice. Self-reliance played an important role, and homeless people were considered as citizens. The excerpt also shows that the State Party tried to find a balance between the rights and duties of property owners on the one hand and the rights and duties of illegal squatters on the other hand. Despite the country's emphasis on private property, evicted persons' (squatters') rights were valued the same as the rights of private property owners. This, though, did not apply to all squatters. The State Party differentiated between legitimate squatters and squatters as rights violators:

> In dealing with the issue of eviction and demolition, one must distinguish between legitimate action by the Government and action that is alleged to be 'excessive exercise of authority'. One must likewise distinguish between the 'legitimate squatters,' those urban poor in need of government assistance for housing, and 'professional squatters' or those who could afford housing but choose the squatting option to avoid paying for housing or, worse, for economic gain. (The Philippines 1994: para. 368)

The Philippines are regarded as a country on the road to social citizenship based on the states' self-description in the States Parties reports. The Committee would, of course, deny that the country entered the road to social citizenship. The Philippines have received urgent appeals from the Committee regarding their eviction policy (2.3).

All in all, states on the road to social citizenship are rare, but are distributed across three regions of the world: the Americas, Europe, and Asia. There are, however, two different paths to turn on the road to social citizenship: the economic path and the non-economic path. When the states used the economic path, they focused on Spaces of Inadequate

Housing and talked about how they tried to integrate people into the society through land titling. They also showed characteristics of globalization of social policy. When they used the non-economic path, they tried to meet at least some rights and fill some of the needs of people affected by inadequate housing, such as food, clothing, health, and education without market participation. States then talked about facilities that offered a wide range of services for homeless people.

7
Conclusion

On a global level, the number of people living in poverty is decreasing (UNDP 2014), but the number of people living in inadequate housing is not (Kothari & Chaudhry 2012). In this book, I have not concentrated on housing policies, but posited the manner of talking about housing as the center of my research interest. For this purpose, the UN human rights served as the point of departure.

I questioned how the ICESCR member states interpreted their obligations concerning the human right to housing and how the Committee accompanied and influenced this reporting. As expected, the ICESCR discourse provided several answers. With respect to the relationship between land policy, social policy, and housing, I discovered some astonishing findings in the scholarly discourse.

During this research, I discovered that housing is an issue of social policy. Housing researchers discuss housing from the perspective of social policy, while social policy researchers seldom examine housing in detail. Looking at housing with a view to social policy usually means to debate the tenure question: homeownership or renting? However, at present, there is still no global consensus on which form of tenure is best. When considering housing satisfaction, the tenure aspect is only one of

M. Kolocek, *The Human Right to Housing in the Face of Land Policy and Social Citizenship*, DOI 10.1007/978-3-319-53489-3_7

many other reasons to determine whether people are satisfied with their housing situation or not. Kemeny (2005: 74) is right when suggesting further research to evaluate the advantages and disadvantages of home-ownership and renting. I do not expect that such research will bring the debate to an end, but may help bring understanding to why the human right to housing does not elicit a single response. This is particularly true when considering housing in a global context. So far, scholars who have regarded housing from a social policy perspective have had Western states in mind (Doling 1997). They considered housing in general and only rarely took inadequate housing into account. If they did, the discussion centered on questions such as how to integrate homeless people back into formal housing markets, or how to prevent them from falling through the cracks of these markets.

The book has evaluated the role of land policy by implementing the human right to housing. I presented two approaches of socio-ecological land policy: land titling and informality. Land titling, understood as 'the allocation of real property rights on land' (Payne et al. 2009: 444), is supposed to wake up the dead capital embodied in people living and working in informal settings (de Soto 2000). The concept of informality contains several positions that suggest other arrangements besides formal property rights contracts. Such positions propose recognizing the 'every-day social contract of informality' (B. Davy & Pellissery 2013) that is based on trust, security, stability, protection, and control. Their message is that the human right to housing can also be implemented outside of formal property markets. My theory about the de-commodification of land use shows what such an approach could look like.

So far, I have not discussed whether housing is more of an economic, social, or even a cultural right. From an economic point of view, land as a commodity could either be a cause or a solution for inadequate housing. The key idea—land policy as social policy—is more than a century old (Damaschke 1918 [1902]; Engels 1942 [1872]: 19). The main difference in comparison with previous centuries, however, is that now researchers, states, UN organs, international NGOs, and other global actors discuss such ideas on a global level. The ICESCR discourse proves that increasing land values are still considered a potential cause or cure for inadequate housing. As the Committee stated, the unbridled speculation in land was one main reason for forced evictions (CESCR 1997: para. 7). The states were adamantly urged to protect people from evictions in the Concluding Observations. The rising degree of attention given to land titling in the States Parties reports shows that land as a commodity was also regarded as a solution. Housing is, however, neither only an economic right nor merely a social right. It is both (and it also has a cultural dimension), and, even more importantly, it is interrelated with other 'typical' economic and social rights. Depending on the form of inadequate housing in question, the states' manner of reporting differed.

The book differentiated between two main forms of inadequate housing: homelessness and Spaces of Inadequate Housing. Homelessness is, in contrast to Spaces of Inadequate Housing, often un-located, meaning that homeless people use different places to sleep and meet other needs. The category of homelessness is complex, and the borders between homelessness and Spaces of Inadequate Housing often blur. Several States Parties (particularly European ones) mentioned night shelters in their reports. The present research cannot evaluate if the mentioned night shelters should be regarded as inadequate, due to security aspects or overcrowding, for instance. In contrast to homelessness, Spaces of Inadequate Housing are, as the name indicates, spaces where people live and meet their needs permanently. These spaces are, inter alia, inadequate because of lack of tenure security, overcrowding, or lack of access to basic amenities. Slums were under the three most frequently mentioned forms of Spaces of Inadequate Housing. A slum dweller in Mumbai, whose every morning begins with the search for a place to defecate, has a lot in common with a homeless person living in Dortmund. There are, however, several differences.

My analysis confirmed these differences in the ICESCR discourse. The ICESCR member states began to recognize (respect) illegal and informal housing structures by responding to Spaces of Inadequate Housing. This recognition usually had an economic background. Land titling or microcredits appeared more often than soup kitchens and other facilities that respond to certain needs of the affected persons. The States Parties also began to increasingly recognize homeless persons as citizens. In scholarly discourses, as well as in the ICESCR discourse, homeless people were usually regarded as people without economic potential, while Spaces of Inadequate Housing were regarded as an economic issue. This could also be proven by comparing the involvement of different actors with each other. Market actors were mentioned more frequently when the states reported Spaces of Inadequate Housing. This applies to self-help/family and supranational actors as well. When states reported homelessness, they most often reported NGO involvement. Homelessness was an issue for supranational actors when states gave an account of street children. Additionally, policies responding to inadequate housing were more often global when states reported Spaces of Inadequate Housing or spoke of street children instead of reporting on adult homeless persons. The World Bank, which is one of the most prominent global actors for both land policy and housing, surprisingly did not have a dominant role concerning inadequate housing in the ICESCR discourse.

The research has shown that housing policies under the umbrella of de-commodification of land use are still in their infancy. I do not state that such a policy always responds better to inadequate housing than an economically-oriented policy (such as land titling). For instance, in media discourse, the rising number of soup kitchens is not usually regarded as an innovative form of social policy, but rather as an indicator of a failed welfare state. Housing is a complex issue. Therefore, different measures are required to respond to any inadequacies. However, land titling, as well as de-commodification of land use, both highlight the significance of land in discourses about the implementation of the human right to housing. At present, millions of people live outside of formal housing markets. Consequently, the states and all other responsible actors should keep in mind that the economic function of land is only one of many others. The states did not explicitly state this, but their way of talking about housing indicates that they have already begun to recognize that in the future, they will not be able to completely fulfill the human right to housing. Of course, no human right can be fulfilled overnight. The language of the ICESCR acknowledges that by using the term of the progressive realization (Article 2, para. 1, ICESCR) of ESC rights. But even this progressive realization becomes increasingly unlikely. The respect-protect-fulfill typology of the Maastricht Guidelines

(1997) is an 'interpretative tool' (Dankwa et al. 1998: 713) that helps to highlight the changing interpretation of housing in the group of other ESC rights. The order of the three verbs respect, protect, and fulfill indicates three consecutive steps toward really implementing ESC rights. The respecting aspects should come first, then states should protect ESC rights, and lastly comes the fulfill aspect. In the context of the interpretation of the human right to housing, I could not prove such an order for the 38 years of reporting in the ICESCR discourse. Instead, it even worked the other way around.

At the beginning of the discourse (first period), the states dominantly reported construction measures. Housing was an issue of mass supply, and the states posited themselves as the key responsible actors. In the second period, the States Parties' and the Committee's attention to housing significantly increased. At the same time, in the Concluding Observations and especially in General Comment No. 7 (CESCR 1997), the Committee criticized the practice of forced evictions that affected people in both developed and developing countries. The States were vehemently urged to protect people from falling into a situation of inadequate housing caused by forced evictions. In the third period, the respect aspect was on the rise. The States Parties regarded people affected by inadequate housing as rights holders rather than right violators. In the States Parties reports, the states were merely one actor of many others involved in housing policies. They discussed inclusion policies and posited equality as an explicit policy goal. Housing was increasingly regarded through global lenses. The rate of mentioning supranational actors, international conferences, and housing policies for refugees or asylum seekers also increased. The growing recognition of informal housing structures could be proven for the ICESCR discourse, scholarly literature, and the international jurisprudence, especially the jurisprudence of the European Court of Human Rights.

By reporting housing, African ICESCR member states were either focused on Spaces of Inadequate Housing or street children, or they reported that they were very poor and needed help. Furthermore, a high percentage of African countries stood out because they gave little detail about inadequate housing. The group of the ICESCR member states that have not submitted any report so far also consists mainly of African countries.

ICESCR member states from Latin America and the Caribbean often intensively reported inadequate housing from an economic point of view. Self-help or family and supranational actors played a more significant role than in other UN regions. They afforded more attention to Spaces of Inadequate Housing than to homelessness.

The European ICESCR member states Finland, France, Greece, the Netherlands, Poland, Spain, and the United Kingdom reported housing in a language of equality, social integration, and individual rights. Usually, they elaborated more on homelessness than on Spaces of Inadequate Housing. A few of them even showed that they entered a road to global social citizenship. Canada and the three countries from Oceania had more in common with the European states than with the states from Asia, Africa, or the Americas.

In Asia, many ICESCR member states gave little, if any, thought to inadequate housing. The Philippines is one of the most interesting states in the ICESCR discourse. The country has been admonished by the Committee regarding forced evictions. Perhaps the Philippines is the best example to demonstrate the gap between what a country reported in its States Parties reports (the discourse level) and how the country actually conducted itself on a policy level. The present research did not evaluate to what extent the states reported the truth or not. It did, however, draw a picture (or rather, considering the tables and diffusion maps, it drew several pictures) of the states' own views of the human right to housing and the policies to respond to it. This picture can be interpreted differently, depending on the perspective.

Optimistically interpreting the research findings means regarding Chapters 5 and 6 as a story of increasing attention to housing. These chapters show that there is undoubtedly a trend toward social citizenship in the States Parties reports. States recognized the complexity of inadequate housing and offered a broad catalog of measures to respond to it. They have already begun to recognize that millions of people are living outside of formal housing markets. Most of them have acknowledged that regarding them as criminals does not solve their housing problems. States Parties worldwide reported many groups of people affected by (or disadvantaged by and vulnerable to) inadequate housing—nationals as well as non-nationals such as asylum seekers or refugees—with a growing degree of detail. Therefore, a trend toward the globalization of social policy could also be proven in the ICESCR discourse. A few states not only entered the road to national social citizenship, but even to global social citizenship. Optimists can regard the ratification of the Optional Protocol to the ICESCR in 2013 as an important discourse event that stands for a growing global awareness of ESC rights. Since the ratification of the Optional Protocol, individuals or groups of individuals have now been able to claim a violation of their rights before the Committee on Economic, Social and Cultural Rights. Optimists could also point out the growing influence of this Committee and its partners (such as the Special Rapporteur on inadequate housing or UNDP), which I proved for the second and the third reporting period. The Committee respected the complexity of housing and responded to each State Party report individually. It showed a critical point of view when states left their people alone as victims of the free forces of the markets. This became particularly visible with respect to privatization policies. The Committee tried to push the states into a more global direction, but criticized them heavily when it saw negative impacts on the most vulnerable. This could be proven by regarding the Committee's critical comments concerning the structural adjustment programs by the World Bank and the IMF. To sum up the optimistic perspective: ESC rights, including the human right to housing, are on the rise, and the ICESCR discourse does not only confirm, but enforces this trend.

Pessimists, however, can also find their arguments. They could stress that, although ESC rights have been on the agenda for many decades, most countries have still not respected, protected, and fulfilled them sufficiently. The ICESCR and its monitoring system could not prevent that more than 1.6 billion people currently live in situations of inadequate housing (Kothari & Chaudhry 2012). Moreover, pessimists could point to the fact that a third of the reporting states gave little, if any, attention to inadequate housing in their reports (and several ICESCR members have not even submitted a report so far). Pessimists could raise the argument that the United States has not yet ratified the Covenant. They could also point out that more than half of the reporting States Parties did not discuss any of the vulnerable and disadvantaged groups in the context of housing. Although the term informal seems to become more prominent, in all, the states have been talking more about illegal settlements instead of informal settlements. So far, only a few states have entered the road to social citizenship. Pessimists could claim that states have, so far, not sufficiently recognized that land is more than just a commodity. They also could claim that in the past, only a few scholars and only a small number of NGOs have paid heed to the contents of the ICESCR discourse at all.

This book dealt with the States Parties' consideration of inadequate housing and analyzed the self-descriptions, not the concrete realization. Every single piece of presented information in the text, the tables, or the diffusion maps cries for further research that answers questions such as 'Have the states reported the truth or not?' or 'Why did a state report this subject and not another one?' The ICESCR discourse is only one of many others. The findings cry for a critical review on other discourse levels and, of course, on the policy level. Further research is possible with the present database. This applies particularly to the comparison section. For instance, I did not go into detail comparing OECD countries with other countries, or countries with different legal or cultural traditions with each other, or former communist countries with Western countries, and so on. Moreover, a comparison between the manner of speaking about housing and other ESC rights, such as the right to food, would be interesting. I am sure that, again, land would play a key role in such a study.

The ICESCR monitoring system has only been recognized by a small number of researchers, human rights experts, and NGOs. The potentials of the ICESCR monitoring system as an appropriate instrument to raise the concerns of people in situations of inadequate housing can only be mobilized once the public attention to the States Parties reports and the Concluding Observations increases. More researchers, NGOs, and other experts should testify whether the states reported the truth or not. The media should play a key role by raising public awareness of the ICESCR discourse. All actual States Parties reports and Concluding Observations are easily available on the Internet. The comparisons between the UN regions have demonstrated that, when speaking about responses to many forms of inadequate housing, the states can also learn a lot from each other. Homelessness is neither a problem of only Western countries nor do Spaces of Inadequate Housing only exist in the global South. Socio-ecological land policy (B. Davy 2009) can respond to many forms of inadequate housing—worldwide.

Housing is a human right. The ICESCR discourse is a valuable data source to examine how this right has been interpreted globally. With more attention from the civil society, the ICESCR discourse could become more than that.

Bibliography

The States Parties reports are listed based on how the states named themselves in the submitted reports. China, for instance, called itself 'People's Republic of China' in the report submitted in 2003 and 'China' in the report submitted in 2010. I did not list all documents from Database FLOOR A (SPR and CO), but instead listed only those that have been quoted in the text.

Ajayi, Miriam; Alexandra Becker; Marina Eismann; Katharina Hoppe; Simone Hoppmann; Manja Laske; Laura Rakow; Sabrina Schmitz; Martin Többen; Camilla Willeke; Inga Marie Wolf & Franziska Zibell (2011) Gemeinsam und für sich. Wohnraum für Obdachlose in Hamburg. Final report at the Faculty of Spatial Planning, TU Dortmund University. Unpublished.

Aldrich, Brian C. & Ranvinder S. Sandhu (1995) The global context of housing poverty. In: Brian C. Aldrich & Ranvinder S. Sandhu (editors): 17–33.

Aldrich, Brian C. & Ranvinder S. Sandhu (editors) (1995) Housing the urban poor. Policy and practice in developing countries. London & New Jersey: Zed Books.

Alston, Philip (1990) U.S. ratification of the Covenant on Economic, Social and Cultural Rights: The need for an entirely new strategy. American Journal of International Law 84 (2): 365–393.

M. Kolocek, *The Human Right to Housing in the Face of Land Policy and Social Citizenship*, DOI 10.1007/978-3-319-53489-3

Alston, Philip (1992) The Committee on Economic, Social and Cultural Rights. In: Philip Alston (editor) The United Nations and human rights. A critical appraisal. Oxford: Clarendon: 473–508.

Alterman, Rachelle (2002) Planning in the face of crisis. Land use, housing and mass immigration in Israel. London & New York: Routledge.

Amore, Kate; Michael Baker & Chapman Philippa Howden (2011) The ETHOS definition and classification of homelessness: An analysis. European Journal of Homelessness 5 (2): 19–37.

Angel, Shlomo (2000) Housing policy matters. A global analysis. Oxford: Oxford University Press.

Arts, Wil & John Gelissen (2002) Three worlds of welfare capitalism or more? A state-of-the-art report. Journal of European Social Policy 12 (2): 137–158.

Avramov, Dragana (1995) Homelessness in the European Union. Social and legal context on housing exclusion. Brussels: FEANTSA.

Baptista, Isabel & Eoin O'Sullivan (2008) The role of the state in developing homeless strategies: Portugal and Ireland in comparative perspective. European Journal of Homelessness 2: 25–43.

Baron, Jane B. (2004) Homelessness as a property problem. The Urban Lawyer 36 (2): 273–288.

Barrientos, Armando (2004) Latin America: Towards a liberal-informal welfare regime. In: Ian Gough & Geoffrey D. Wood (editors): 121–168.

Benjaminsen, Lars & Evelin Dyp (2008) The effectiveness of homeless policies – Variations among the Scandinavian countries. European Journal of Homelessness 2: 45–67.

Berger, Peter L. & Thomas Luckmann (1967) The social construction of reality. A treatise in the sociology of knowledge. Garden City, NY: Doubleday.

Bevan, Philippa (2004) The dynamics of African in/security regimes. In: Ian Gough & Geoffrey D. Wood (editors): 202–252.

Bierre, Sarah; Philippa Howden-Chapman & Louise Signal (2010) 'Ma and Pa' landlords and the 'risky' tenant: Discourses in the New Zealand private rental sector. Housing Studies 25 (1): 21–38.

Busch-Geertsema, Volker (2014) Housing First Europe. Results of a European social experimentation project. European Journal of Homelessness 8 (1): 13–28.

Busch-Geertsema, Volker & Ingrid Sahlin (2007) The role of hostels and temporary accommodation. European Journal of Homelessness 1: 67–93.

Buschmann, Luise (2010) Die Entstehung des Menschenrechts auf soziale Grundsicherung im International Covenant on Economic, Social and Cultural Rights. Working Paper No. 7. Online: www.floorgroup.de. (Accessed December 2014).

Buschmann, Luise (2013) Das Menschenrecht auf soziale Grundsicherung aus Art. 9 und Art. 11 ICESCR. Münster: agenda.

Campbell, Patricia (2013) Collateral damage? Transforming subprime slum dwellers into homeowners. Housing Studies 28 (3): 453–472.

Castles, Francis G. (1998) The really big trade-off: Homeownership and the welfare state in the new world and in the old. Acta Politica 33 (1): 5–20.

Castles, Francis G. & Francisco H. G. Ferreira (1996) Home ownership and the welfare state: Is Southern Europe different? South European Society & Politics 1 (2): 163–185.

CESCR (Committee on Economic, Social and Cultural Rights) (1986) General Guidelines regarding the form and contents of reports to be submitted by States Parties under Articles 16 and 17 of the International Covenant on Economic, Social and Cultural Rights. E/C.12/1987/2.

CESCR (1989) General Comment No. 1: Reporting by States Parties. In: CESCR (1989) Report on the third session (6–24 February 1989). E/C.12/1989/5. Annex III, 87–89.

CESCR (1990a) General Comment No. 2. International technical assistance measures (Art. 22 of the Covenant). In: CESCR (1990) Report on the fourth session (15 January–2 February 1990). E/C.12/1990/3. Annex III: 86–89.

CESCR (1990b) General Comment No. 3. The nature of States Parties obligation (Art. 2, para. 1, of the Covenant). In: CESCR (1991) Report on the fifth session (26 November–14 December 1990). E/C.12/1990/8. Annex III: 83–87.

CESCR (1991a) General Comment No. 4. The right to adequate housing (Art. 11, para. 1, of the Covenant). In: CESCR (1992) Report on the sixth session. E/C.12/1991/4. Annex III: 114–120.

CESCR (1991b) Revised General Guidelines regarding the form and contents of reports to be submitted by States Parties under Articles 16 and 17 of the International Covenant on Economic, Social and Cultural Rights. E/C.12/1991/1.

CESCR (1993) Rules of Procedure of the Committee. E/C.12/1990/4/Rev.1.

CESCR (1994) Report on the eight and ninth sessions (10–18 May 1993, 22 November–10 December 1993). E/C.12/1993/19.

CESCR (1995) General Comment No. 6. The economic, social and cultural rights of older persons. In: CESCR (1995) Report on the tenth and eleventh sessions (2–20 May 1994, 21 November–9 December 1994). E/C.12/1994/20. Annex IV, 99–109.

CESCR (1997) General Comment No. 7. The right to adequate housing (Art. 11, para. 1, of the Covenant). Forced evictions. E/C.12/1997/4.

CESCR (2002a) General Comment No. 15. The right to water (Arts. 11 and 12 of the Covenant). E/C.12/2002/11.

CESCR (2002b) Statement of the Committee on Economic, Social and Cultural Rights to the special session of the General Assembly for an overall review and appraisal of the implementation of the decisions taken at the United Nations Conference on Human Settlements (Habitat II) (New York, 6–8 June 2001). In: CESCR (2002) Report of the twenty-fifth, twenty sixth and twenty-seventh sessions (23 April–11 May 2001, 13–31 August 2001, 12–30 November 2001). E/C.12/2001/17. Annex XI: 209–210.

CESCR (2006) General Comment No. 18. The right to work. E/C.12/GC/18.

CESCR (2008) General Comment No. 19. The right to social security (Art. 9). E/C.12/GC/19.

CESCR (2009) Guidelines on treaty-specific documents to be submitted by States Parties under Articles 16 and 17 of the International Covenant on Economic, Social and Cultural Rights. E/C.12/2008/2.

Clapham, David (2012) Social policy approaches to housing research. In: David Clapham (editor): 163–173.

Clapham, David (editor) (2012) The SAGE handbook of housing studies. Los Angeles: Sage.

COHRE (Centre for Housing Rights and Evictions) (2002): Enforcing housing rights in the Americas. Pursing housing rights claims within the Inter-American system of human rights. Geneva: COHRE.

Commission on Human Settlements (1999) Guidelines on practical aspects in the realization of the human right to adequate housing, including the formulation of the United Nations housing rights programme. HS/C/17/INF/6.

Council of Europe (2012) Descriptive Glossary of terms relating to Roma issues. 18 May 2012. Online: http://a.cs.coe.int/team20/cahrom/documents/Glossary%20Roma%20EN%20version%2018%20May%202012.pdf. (Accessed November 2016).

Craven, Matthew (2001) The UN Committee on Economic, Social and Cultural Rights. In: Asbjørn Eide, Catarina Krause & Allan Rosas (editors): 455–472.

Craven, Matthew (2002) The International Covenant on Economic, Social and Cultural Rights. A perspective on its development. Oxford: Clarendon Press.

Cumming, Susanna & Tsuyoshi Ono (1997) Discourse and grammar. In: Teun A. van Dijk (editor) (1997b): 112–137.

Czasny, Karl; Heidrun Feigefeld; Jürgen Hajek; Peter Moser & Eva Stocker (2008) Wohnzufriedenheit und Wohnbedingungen in Österreich im europäischen Vergleich. Wien: SRZ.

Daly, Gerald (1996) Homeless. Policies, strategies, and lives on the street. London & New York: Routledge.

Damaschke, Adolf (1918 [1902]) Die Bodenreform. Grundsätzliches und Geschichtliches zur Erkenntnis und Überwindung der sozialen Not. 15th edition. Jena: Gustav Fischer.

Dankwa, Victor; Cees Flinterman & Scott Leckie (1998) Commentary to the Maastricht Guidelines on violations of economic, social and cultural rights. Human Rights Quarterly 20: 705–730.

Davis, Mike (2006) Planet of slums. London & New York: Verso.

Davy, Benjamin (2000) Welches Eigentum braucht die Stadt? vhw Forum Wohneigentum 8 (August): 288–294.

Davy, Benjamin (2005) Bodenpolitik. In: Akademie für Raumforschung und Landesplanung (editor) Handwörterbuch der Raumordnung. 4th edition. Hannover: ARL: 117–125.

Davy, Benjamin (2009) The poor and the land: Poverty, property, planning. Town Planning Review 80 (3): 227–265.

Davy, Benjamin (2012) Land policy. Planning and the spatial consequences of property. Farnham, UK & Burlington, VT: Ashgate.

Davy, Benjamin (2014) Raumplanung und die Politik der Würde. In: Wolfgang Blaas, Johann Bröthaler, Michael Getzner & Gerlinde Gutheil-Knopp-Kirchwald (editors) Perspektiven der staatlichen Aufgabenerfüllung: Zwischen budgetärer Knappheit und integrativem Anspruch. Für Wilfried Schönbäck zum 70. Geburtstag. Wien: Verlag Österreich: 51–76.

Davy, Benjamin & Sony Pellissery (2013) The citizenship promise (un)fulfilled: The right to housing in informal settings. In: Benjamin Davy, Ulrike Davy & Lutz Leisering (guest editors) Exploring global social citizenship: Human rights perspectives. International Journal of Social Welfare 22 (Supplement 1): S68–S84.

Davy, Benjamin; Ulrike Davy & Lutz Leisering (2013) The global, the social and rights. New perspectives on social citizenship. In: Benjamin Davy, Ulrike Davy & Lutz Leisering (guest editors) Exploring global social citizenship: Human rights perspectives. International Journal of Social Welfare 22 (Supplement 1): S1–S14.

Davy, Benjamin; Ulrike Davy & Lutz Leisering (guest editors) (2013) Exploring global social citizenship: Human rights perspectives. International Journal of Social Welfare 22 (Supplement 1).

Davy, Ulrike (2013) Social citizenship going international: Changes in the reading of UN-sponsored economic and social rights. In: Benjamin Davy, Ulrike Davy & Lutz Leisering (guest editors) Exploring global social citizenship: Human rights perspectives. International Journal of Social Welfare 22 (Supplement 1): S15–S31.

Davy, Ulrike (2014) How human rights shape social citizenship: On citizenship and the understanding of economic and social rights. Washington University Global Studies Law Review 13 (2): 201–263.

Davy, Ulrike (2015) Der Universalismus der Allgemeinen Erklärung der Menschenrechte. Die Arbeit am Konsens, 1946–1948. In: Bettina Heinz & Britta Leisering (editors) Menschenrechte in der Weltgesellschaft. Deutungswandel und Wirkungsweise eines globalen Leitwerts. Frankfurt & New York: Campus Verlag: 198–235.

de Janvry, Alain; Gustavo Gordillo & Jean-Philppe Platteau (editors) (2001) Access to land, rural poverty, and public action. Oxford: University Press.

de Soto, Hernando (2000) The mystery of capital. Why capitalism triumphs in the west and fails everywhere else. New York: Basic Books.

Deacon, Bob (2007) Global social policy & governance. Los Angeles: Sage Publications.

Deacon, Bob (2010) Theorising global social policy. Globale Sozialpolitikforschung und Weltgesellschaftsforschung. Zeitschrift für Sozialreform 56 (2): 151–172.

Dean, Hartley (2006) Social policy. Cambridge & Malden, MA: Polity.

Dean, Hartley (2007) Social policy and human rights: Re-thinking the engagement. Social Policy & Society 7 (1): 1–12.

Dean, Hartley (2013) The translation of needs into rights: Reconceptualising social citizenship as a global phenomenon. In: Benjamin Davy, Ulrike Davy & Lutz Leisering (guest editors): S32–S49.

Deaton, Angus (2013) The great escape. Health, wealth, and the origins of inequality. Princeton: Princeton University Press.

Deininger, Klaus (2003) Land policies for growth and poverty reduction. Washington, DC: World Bank & Oxford University Press.

Deininger, Klaus & Hans Binswanger (2001) The evolution of World Bank's land policy. In: Alain de Janvry, Gustavo Gordillo & Jean-Philppe Platteau (editors): 406–440.

Dixon, John A.; Don H. Foster; Kevin Durrheim & Lindy Wilbraham (1994) Discourse and the politics of space in South Africa: The 'squatter crisis'. Discourse & Society 5 (3): 277–296.

Dodson, Jago (2007) Government discourse and housing. Aldershot, UK & Burlington, VT: Ashgate.

Doling, John (1997) Comparative housing policy. Government and housing in advanced industrialized countries. Basingstoke & New York: Palgrave Macmillan.

Doyle, Arthur Conan (2003 [1891]) Sherlock Holmes. The complete novels and stories. New York: Bantam Classics.

Dzudzek, Iris; Georg Glasze; Annika Mattissek & Henning Schirmel (2009) Verfahren der lexikometrischen Analyse von Textkorpora. In: Georg Glasze & Annika Mattissek (editors): 233–260.

ECE (United Nations Economic Commission for Europe) (2009) Self-made cities. In search for sustainable solutions for informal settlements in the United Economic Commission for Europe region. New York & Geneva: United Nations Publications. ECE/HBP/155.

ECOSOC (Economic and Social Council) (1985) ECOSOC resolution 1985/17. Review of the composition, organization and administrative arrangements of the sessional working group of governmental experts on the implementation of the International Covenant on Economic, Social and Cultural Rights.

ECOSOC (1995) ECOSOC resolution 1995/39.

ECOSOC (2001) Report of the Special Rapporteur on adequate housing as a component of the right to an adequate standard of living, Mr. Miloon Kothari, submitted pursuant to Commission resolution 2000/9. E/CN.4/2001/51.

Edgar, Bill; Joe Doherty & Henk Meert (2002) Access to housing. Homelessness and vulnerability in Europe. Bristol: Policy Press.

Eide, Asbjørn (2001) The Right to an adequate standard of living including the right to food. In: Asbjørn Eide, Catarina Krause & Allan Rosas (editors): 133–148.

Eide, Asbjørn (2010) Freedom from need: The universal right to an adequate standard of living – origins, obstacles and prospects. Scandinavian Studies in Law 55: 157–180.

Eide, Asbjørn, Catarina Krause & Allan Rosas (editors) (2001) Economic, social and cultural rights. A textbook. 2nd edition. Dordrecht, Boston & London: Martinus Nijhoff Publishers.

Elsinga, Marja & Joris Hoekstra (2005) Homeownership and housing satisfaction. Journal of Housing and the Built Environment 20: 401–424.

Engels, Frederick (1942 [1872]) The housing question. 2nd edition. London: Lawrence and Wishart Ltd.

Esping-Andersen, Gøsta (2011 [1990]) The three worlds of welfare capitalism. Cambridge: Polity Press.

EUROFOUND (European Foundation for the Improvement of Living and Working Conditions) (2009) Second European quality of life survey. Overview. Luxembourg: Office for Official Publications of the European Communities.

Every, Danielle & Martha Augoustinos (2007) Constructions of racism in the Australian parliamentary debates on asylum seekers. Discourse & Society 18 (4): 411–436.

FEANTSA (2005) European typology of homelessness and housing exclusion. Online: http://www.feantsa.org/spip.php?article120&lang=en. (Accessed December 2014).

Felice, William F. (1999) The viability of the United Nations approach to economic and social rights in a globalized economy. International Affairs 75 (3): 563–598.

Fernandes, Edésio (2011) Regularization of informal settlements in Latin America. Cambridge, MA: Lincoln Institute of Land Policy.

Ferreira, Francisco H. G. & Lousie C. Keely. (2000) The World Bank and structural adjustment: Lessons from the 1980s. In: Christopher L. Gilbert & David Vines (editors): 159–195.

FIDH (International Federation for Human Rights) (2010) Practical Guide. The African Court on Human and Peoples' Rights towards the African Court of Justice and Human Rights. Paris: FIDH.

Flinterman, Cees (1997) The Maastricht Guidelines on violations of economic, social and cultural rights. Netherlands Quarterly of Human Rights 15 (2): 244–252.

Fopp, Rodney (2009) Metaphors in homelessness discourse and research: Exploring 'pathways', 'careers' and 'safety nets'. Housing, Theory and Society 26 (4): 271–291.

Forte, James A. (2002) Not in my social world: A cultural analysis of media representations, contested spaces, and sympathy for the homeless. Journal of Sociology and Social Welfare 29: 131–158.

Foucault, Michel (2002 [1969]) Archaeology of knowledge. London & New York: Routledge.

Fox O'Mahony, Lorna & James A. Sweeney (editors) (2011) The idea of home in law. Displacement and dispossession. Farnham, UK & Burlington, VT: Ashgate.

Fragemann Mathis (2014) PfandsammlerInnen in öffentlich zugänglichen Räumen – Anfeindung und Verdrängung oder Unterstützung und Integration? Diploma thesis at the Faculty of Spatial Planning, TU Dortmund University. Unpublished.

Friese, Susanne (2013) ATLAS.ti 7 User Guide and Reference. Berlin: ATLAS.ti Scientific Software Development GmbH. Online: http://atlasti.com/wp-content/uploads/2014/05/atlasti_v7_manual_201312.pdf?q=/uploads/media/atlasti_v7_manual_201312.pdf. (Accessed December 2014).

Friese, Susanne (2014) Qualitative data analysis with ATLAS.ti. 2nd edition. London: Sage.

Friese, Susanne & Thomas Ringmayr (editors) (2014) ATLAS.ti user conference 2013: Fostering dialog on qualitative methods. Berlin: Universitätsverlag der TU Berlin. Online: http://opus4.kobv.de/opus4-tuberlin/frontdoor/index/index/docId/5157. (Accessed December 2014).

Fritzsche, Karl Peter (2009) Menschenrechte. Eine Einführung mit Dokumenten. 2nd edition. Paderborn et al.: Schöningh.

Galster, George (2012) Neighborhoods and their role in creating and changing housing. In: David Clapham (editor): 84–106.

Gilbert, Alan (2002) On the mystery of capital and the myths of Hernando de Soto. What difference does legal title make? International Development Planning Review 24 (1): 1–19.

Gilbert, Alan (2007) The return of the slum: Does language matter? International Journal of Urban and Regional Research 31 (4): 697–713.

Gilbert, Christopher L. & David Vines (editors) (2000) The World Bank. Structure and policies. Cambridge: Cambridge University Press.

Gilbert, Christopher L. & David Vines (2000) The World Bank: An overview of some major issues. In: Christopher L. Gilbert & David Vines (editors): 10–36.

Gill, Ann M. & Karen Whedbee (1997) Rhetoric. In: Teun A. van Dijk (editor) (1997b): 157–184.

Glaser, Barney G. & Anselm L. Strauss (2008 [1967]) The discovery of grounded theory. Strategies for qualitative research. 3rd edition. New Brunswick, NJ: Aldine Transaction.

Glasze, Georg & Annika Mattissek (editors) (2009) Handbuch Diskurs und Raum. Theorien und Methoden für die Humangeographie sowie die sozial- und kulturwissenschaftliche Raumforschung. Bielefeld: Transcript.

Gorsler, Daniela (2002) Informelle räumliche Planung. Stand der aktuellen Forschung und Forschungsbedarf. Hannover: Verlag der ARL.

Gough, Ian (2004a) Welfare regimes in development contexts: A global and regional analysis. In: Ian Gough & Geoffrey D. Wood (editors) Insecurity and welfare regimes in Asia, Africa, and Latin America. Social policy in development contexts. Cambridge: Cambridge University Press: 15–48.

Gough, Ian (2004b) East Asia: The limits of productivist regimes. In: Ian Gough & Geoffrey D. Wood (editors): 169–201.

Gough, Ian & Geoffrey D. Wood (editors) (2004) Insecurity and welfare regimes in Asia, Africa, and Latin America. Social policy in development contexts. Cambridge: Cambridge University Press.

Günzel, Marian (2016) »Was im Dunkeln liegt, kann man nicht sehen« – Eine Spurensuche nach diskurstheoretischen Impulsen in der Raumplanung. Online: http://www.planung-neu-denken.de/images/stories/pnd/dokumente/1_2016/gnzel.pdf. (Accessed November 2016).

Hamburg (Freie und Hansestadt Hamburg, Behörde für Arbeit, Soziales, Familie und Integration) (2013) Das soziale Hilfesystem für wohnungslose Menschen. Hamburg: Eigendruck.

Harris, D. J. (2009) Collective complaints under the European Social Charta: Encouraging progress? In: Kaiyan Homi Kaikobad & Michael Bohlander (editors) International law and power. Perspectives on legal order and justice. Essays in Honour of Colin Warbrick. Leiden: Brill: 3–24

Hartmann, Thomas (2011) Clumsy floodplains. Responsive land policy for extreme floods. Farnham, UK & Burlington, VT: Ashgate.

Harvey, David (2013) Rebel cities: From the right to the city to the urban revolution. London & New York: Verso.

Healey, Patsy (1992) Planning through debate. The communicative turn in planning theory. Town Planning Review 63 (2): 143–162.

Henkin, Louis (1979) International human rights as 'rights'. Cardozo Law Review 1: 425–447.

Henkin, Louis (1990) The age of rights. New York: Columbia University Press.

Himpe, Nina-Claire (2013) „Das Soziale" in der Allgemeinen Erklärung der Menschenrechte. Ein Kompromiss zwischen Staaten der Welt. Working Paper No. 17. Online: www.floorgroup.de. (Accessed December 2014).

Hoekstra, Joris (2003) Housing and the welfare state in the Netherlands. An application of Esping-Andersen's typology. Housing, Theory and Society 20 (2): 58–71.

Hoekstra, Joris (2005) Is there a connection between welfare state regime and dwelling type? An exploratory statistical analysis. Housing Studies 20 (3): 475–495.

Hoekstra, Joris (2010) Divergence in European welfare and housing systems. Amsterdam: IOS Press.

Hradecký, Ilja (2008) Building capacity of homeless services in the Czech Republic. European Journal of Homelessness 2: 177–190.

Huckin, Thomas (2002) Textual silence and the discourse of homelessness. Discourse & Society 13 (3): 347–372.

ICESCR-OP (2008) Resolution adopted by the General Assembly on 10 December 2008. Optional Protocol to the International Covenant on Economic, Social and Cultural Rights. A/RES/63/117.

Isin, Engin F. & Bryan S. Turner (editors) (2002) Handbook of citizenship studies. London & Thousand Oaks, CA: Sage.

Jacobs, Harvey M. (2013) Private property and human rights: A mismatch in the 21st century? In: Benjamin Davy, Ulrike Davy & Lutz Leisering (guest editors): S85–S101.

Keller, Reiner (2009) Müll – die gesellschaftliche Konstruktion des Wertvollen. Die öffentliche Diskussion über Abfall in Deutschland und Frankreich. Wiesbaden: VS Verlag für Sozialwissenschaften.

Keller, Reiner (2011a) Diskursforschung. Eine Einführung für SozialwissenschaftlerInnen. Wiesbaden: VS Verlag für Sozialwissenschaften.

Keller, Reiner (2011b) Wissenssoziologische Diskursanalyse. Grundlegung eines Forschungsprogramms. 3rd edition. Wiesbaden: VS Verlag für Sozialwissenschaften.

Keller, Reiner (2011c) The sociology of knowledge approach to discourse (SKAD). Human Studies 34 (1): 43–65.

Keller, Reiner (2013) Doing discourse research. An introduction for social scientists. London: SAGE Publications Ltd.

Keller, Reiner; Andreas Hirseland; Werner Schneider & Willy Viehöver (editors) (2006) Handbuch Sozialwissenschaftliche Diskursanalyse. Band 1: Theorien und Methoden. 2nd edition. Wiesbaden: VS Verlag für Sozialwissenschaften.

Keller, Reiner; Andreas Hirseland; Werner Schneider & Willy Viehöver (editors) (2008) Handbuch Sozialwissenschaftliche Diskursforschung. Band 2: Forschungspraxis. 3rd edition. Wiesbaden: VS Verlag für Sozialwissenschaften.

Kemeny, Jim (1980) Home ownership and privatization. International Journal of Urban and Regional Research 4 (372–388).

Kemeny, Jim (1981) The myth of home-ownership. Private versus public choices in housing tenure. London & Boston: Routledge & Kegan Paul.

Kemeny, Jim (2001) Comparative housing and welfare: Theorising the relationship. Journal of Housing and the Built Environment 16 (1): 53–70.

Kemeny, Jim (2005) 'The really big trade-off' between home ownership and welfare: Castles' evaluation of the 1980 thesis, and a reformulation 25 years on. Housing, Theory and Society 22 (2): 59–75.

Kenna, Padraic (2008) Globalization and housing rights. Indian Journal of Global Legal Studies 15 (2): 396–469.

Kenna, Padraic (editor) (2014) Contemporary housing issues in a globalized world. Farnham, UK & Burlington, VT: Ashgate.

Kenna, Padraic & Mark Jordan (2014) Housing rights in Europe: The Council of Europe leads the way. In: Padraic Kenna (editor): 115–140.

Kolocek, Michael (2009) Der Shopping Center Diskurs. Diploma thesis at the Faculty of Spatial Planning, TU Dortmund University. Unpublished.

Kolocek, Michael (2012) Land policy and the social floor to housing. The case of Latin America and Africa. Working Paper No. 15. Online: www.floorgroup.de. (Accessed December 2014).

Kolocek, Michael (2013) The human right to housing in the 27 member states of the European Union. European Journal of Homelessness 7 (1): 135–154.

Kolocek, Michael (2014) The human right to housing: Using ATLAS.ti to combine qualitative and quantitative methods to analyze global discourses. In: Susanne Friese & Thomas Ringmayr (editors). Online: opus4.kobv.de/opus4-tuberlin/files/5896/04_kolocek_second_edition_5896.pdf. (Accessed December 2014).

Kolocek, Michael (2015) Das Menschenrecht auf Wohnen. Interpretationen im globalen Kontext. RaumPlanung 182 (6): 10–15.

Kothari, Millon & Shivani Chaudrhy (2012) Housing, land and sustainable development. In: Social Watch Report. The right to a future. 38–41. Online: http://www.socialwatch.org/sites/default/files/Housing2012_eng.pdf. (Accessed November 2016).

Kurz, Karin (2004) Home ownership and social inequality in West Germany. In: Karin Kurz & Hans-Peter Blossfeld (editors) Home ownership and social inequality in comparative perspective. Stanford: Stanford University Press: 21–60.

Leckie, Scott (2001) The human right to adequate housing. In: Asbjørn Eide, Catarina Krause & Allan Rosas (editors) Economic, social and cultural rights. A textbook. 2nd edition. Dordrecht, Boston & London: Martinus Nijhoff Publishers: 149–168.

Lefebvre, Henri (1968) Le droit à la ville. Paris: Anthropos.

Legewie, Heiner (2014) ATLAS.ti – How it all began. (A Grandfather's Perspective). In: Susanne Friese & Thomas Ringmayr (editors). Online: http://nbn-resolving.de/urn:nbn:de:kobv:83-opus4-44140. (Accessed December 2014).

Leisering, Lutz & Armando Barrientos (2013) Social citizenship for the global poor? The worldwide spread of social assistance. In: Benjamin Davy, Ulrike Davy & Lutz Leisering (guest editors): S50–S67.

Leisering, Lutz; Benjamin Davy & Ulrike Davy (2015) The politics of recognition. Changing understandings of human rights, social development and land rights as normative foundation of global social policy. In: Frauke Lachenmann, Tilmann J. Röder & Rüdiger Wolfrum (editors) Max Planck Yearbook of United Nations Law 18. Leiden & Boston: Brill: 565–600.

Limburg Principles (1987) The Limburg Principles on the implementation of the International Covenant on Economic, Social and Cultural Rights. Human Rights Quarterly 9: 122–135.

Lynch, Philip & Jacqueline Cole (2003) Homelessness and human rights: Regarding and responding to homelessness as a human rights violation. Melbourne Journal of International Law 4 (1): 139–176.

Lynn, Nick & Susan Lea (2003) 'A phantom menace and the new apartheid': The social construction of asylum-seekers in the United Kingdom. Discourse & Society 14 (4): 425–452.

Maastricht Guidelines (1997) Maastricht Guidelines on violations of violations of economic, social and cultural rights. Human Rights Quarterly 20 (3): 691–705.

Mahon, Claire (2008) Progress at the front: The draft Optional Protocol to the International Covenant on Economic Social and Cultural Rights. Human Rights Law Review 8 (4): 617–646.

Main, Thomas (1998) How to think about homelessness: Balancing structural and individual causes. Journal of Social Distress and the Homeless 7 (1): 41–54.

Marangos, John (2008) Thomas Paine (1737–1809) and Thomas Spence (1750–1814) on land ownership, land taxes and the provision of citizens' dividend. International Journal of Social Economics, 35 (5): 313–325.

Marshall, Thomas H. (1950) Citizenship and social class and other essays. Cambridge: Cambridge University Press.

Marshall, Thomas H. (1965) The right to welfare. The Sociological Review 13 (3): 261–272.

Mattissek, Annika (2008) Die neoliberale Stadt. Diskursive Repräsentationen im Stadtmarketing deutscher Großstädte. Bielefeld: Transcript.

Matznetter, Walter (2002) Social housing in a conservative welfare state: Austria as an example. Urban Studies 39 (2): 265–282.

Matznetter, Walter & Alexis Mundt (2012) Housing and welfare regimes. In: David Clapham (editor) The SAGE handbook of housing studies. Los Angeles: Sage: 274–294.

Mayring, Philipp (2000) Qualitative content analysis. Forum: Qualitative Social Research 1 (2): Art. 2.

McNaughton Nicholls, Carol (2010) Housing, homelessness and capabilities. Housing, Theory and Society 27 (1): 23–41.

Meert, Henk; Pedro Cabrera; Ivan Christensen; Inger Koch-Nielsen; Roland Maas & Elisabeth Maurel (2004) The changing profiles of homeless people: Homelessness in the written press: A discourse analysis. Online: http://www.feantsaresearch.org/IMG/pdf/2004_homelessness_in_the_written_press.pdf. (Accessed December 2014).

Mey, Günter & Katja Mruck (editors) (2011) Grounded theory reader. 2nd edition. Wiesbaden: VS Verlag für Sozialwissenschaften.

Milbourne, Paul & Paul Cloke (editors) (2006) International perspectives on rural homelessness. London & New York: Routledge.

Moser, Sebastian J. (2014) Pfandsammler. Erkundungen einer urbanen Sozialfigur. Hamburg: Hamburger Edition, HIS.

Neale, Joanne (1997) Homelessness and theory reconsidered. Housing Studies 12 (1): 47–61.

Neuwirth, Robert (2006) Shadow cities. A billion squatters, a new urban world. New York: Routledge.

Neuwirth, Robert (2011) Stealth of nations. The global rise of informal economy. New York: Pantheon Books.

Noorbakhsk, Farhad & Shadan Noorbakhsh (2006) The effects of compliance with structural adjustment programmes on human development in sub-Saharan Africa. In: Alberto Paloni & Maurizio Zanardi (editors) The IMF, World Bank and policy reform. London & New York: Routledge: 143–161.

Nussbaum, Martha C. (2011) Creating capabilities. The human development approach. Cambridge, MA & London: Harvard University Press.

Oestreich, Joel E. (2004) The human rights responsibilities of the World Bank: A business paradigm. Global Social Policy 4 (1): 55–76.

O'Flaherty, Brendan (1998) Making room. The economics of homelessness. Cambridge & London: Harvard University Press.

Olsson, Lars-Erik & Marie Nordfeldt (2008) Homelessness and the tertiary welfare system in Sweden – The role of the welfare state and non-profit sector. European Journal of Homelessness 2: 157–173.

Oren, Michelle, Rachelle Alterman & Yaffa Zilbershats (2014) Housing rights in constitutional legislation: A conceptual classification. In: Padraic Kenna (editor) 141–158.

Ostrom, Elinor (2001) The puzzle of counterproductive property rights reforms: A conceptual analysis. In: Alain de Janvry, Gustavo Gordillo & Jean-Philppe Platteau (editors): 129–150.

O'Sullivan, Eoin & Pascal de Decker (2007) Regulating the private rental housing market in Europe. European Journal of Homelessness 1: 95–117.

Oxley, Michael (2004) Economics, planning and housing. Basingstoke & New York: Palgrave Macmillan.

Paine, Thomas (1903 [1797]) Agrarian justice, being a plan for meliorating the condition of man. 2nd edition. London: Adlard & Parsons.

Payne, Geoffrey (2001) Urban land tenure policy options: Title or rights? Habitat International 25: 415–429.

Payne, Geoffrey; Alain Durand-Lasserve & Carole Rakodi (2009) The limits of land titling and home ownership. Environment and Urbanization 21 (2): 443–462.

Penner, Maurice & Susan Penner (1989) Visual ideologies of street homeless: Comparing editorail cartoons to fieldwork observations. Visual Sociology Review (2): 99–106.

Piccard, Ann M. (2010) The United States' failure to ratify the International Covenant on Economic, Social and Cultural Rights: Must the poor be always with us? The Scholar: St. Mary's Law Review on Minority Issues 13 (2): 231–272.

Piketty, Thomas (2014) Capital in the twenty-first century. Cambridge, MA & London: Harvard University Press.

Ploeger, Hendrick D. & Daniëlle A. Groetelaers (2007) The importance of the fundamental right to property for the practice of planning. An introduction to the case law of the European Court of Human Rights on Article 1, Protocol 1. European Planning Studies 15 (10): 1424–1438.

Pogge, Thomas W. (2007) World poverty and human rights. Cosmopolitan responsibilities and reforms. Cambridge: Polity.

Polanyi, Karl (2001 [1944]) The great transformation. The political and economic origins of our time. 2nd edition. Boston, MA: Beacon Press.

Pugh, Cedric (1991) Housing policies and the role of the World Bank. Habitat International 15 (1/2): 275–298.

Pugh, Cedric (1995) The role of the World Bank in housing. In: Brian C. Aldrich & Ranvinder S. Sandhu (editors) Housing the urban poor. Policy and practice in developing countries. London & New Jersey: Zed Books: 34–92.

Ronald, Richard (2008) The ideology of home ownership. Homeowner societies and the role of housing. Basingstoke & New York: Palgrave Macmillan.

Rosas, Allan (2001) The right of self-determination. In: Asbjørn Eide, Catarina Krause & Allan Rosas (editors): 111–118.

Rosas, Allan & Martin Scheinin (2001) Implementation mechanisms and remedies. In: Asbjørn Eide, Catarina Krause & Allan Rosas (editors): 425–452.

Roy, Ananya (2004) Transnational trespassings: The geopolitics of urban informality. In: Ananya Roy & Nezar Alsayyad (editors) Urban informality. Transnational perspective from the Middle East, Latin America, and South Asia. Lanham: Lexington Books: 289–317.

Roy, Ananya (2005) Urban informality. Toward an epistemology of planning. Journal of American Planning Association 71 (2): 147–158.

Roy, Ananya (2010) Poverty capital. Microfinance and the making of development. New York: Routledge.

Saul, Ben; David Kinley & Jacqueline Mowbray (2014) The International Covenant on Economic, Social and Cultural Rights. Commentary, cases, and materials. First edition. Oxford: Oxford University Press.

Scheinin, Martin (2006) The proposed Optional Protocol to the Covenant on Economic, Social and Cultural Rights: A blueprint for UN human rights treaty body reform – Without amending the existing treaties. Human Rights Law Review 6 (1): 131–142.

Schneider, Friedrich & Andreas Buehn (2009) Shadow economies and corruption all over the world: Revised estimates for 120 countries. Economics: The open Access, Open Assessment E-Journal (Version 2, October): 1–53. Online: http://www.economics-ejournal.org/economics/journalarticles/2007-9. (Accessed December 2014).

Smith, Nick (2009) Commodification in law: Ideologies, intractabilities, and hyperboles. Continental Philosophical Review 42: 101–129.

Somers, Margaret R. (2008) Genealogies of citizenship. Markets, statelessness, and the right to have rights. Cambridge, New York: Cambridge University Press.

Springer, Sabine (2000) Homelessness: A proposal for a global definition and classification. Habitat International 24: 475–484.

Stamsø, Mary-Ann (2010) Housing and welfare policy – Changing relations? A cross-national comparison. Housing, Theory and Society 27 (1): 64–75.

Steinmeier, Frank-Walter (1992) Bürger ohne Obdach. Zwischen Pflicht zur Unterkunft und Recht auf Wohnraum. Bielefeld: VSH Verlag Soziale Hilfe.

Stephens, Mark & Suzanne Fitzpatrick (2007) Welfare regimes, housing systems and homelessness: How are they linked? European Journal of Homelessness 1: 201–212.

Strauss, Margot & Sandra Liebenberg (2014) Contested spaces: Housing rights and evictions in post-apartheid South Africa. Planning Theory 13 (4): 429–448.

Swift, Jonathan (1729) A modest proposal. For preventing the children of poor people in Ireland from being a burden to their parents or country, and for making them beneficial to the public. Online: http://www.secret-satire-society.org/wp-content/uploads/2015/11/Jonathan-Swift-A-Modest-Proposal.pdf. (Accessed November 2016).

Tipple, Graham & Suzanne Speak (2009) The hidden millions. Homelessness in developing countries. London & New York: Routledge.

Torck, Danièle (2001) Voices of homeless people in street newspapers: A cross-cultural exploration. Discourse & Society 12 (3): 371–391.

Torgersen, Ulf (1987) Housing: The wobbly pillar under the welfare state. In: Bengt Turner, Jim Kemeny & Lennart Lundqvist (editors) Between state and market. Housing in the post-industrial era. Stockholm: Almqvist & Wiksell: 116–127.

Turner, Bryan S. (1990) Outline of a theory of citizenship. Sociology 24 (2): 189–217.

UN (United Nations) (1977) Resolution adopted by the General Assembly. 32/162. Institutional arrangements for international cooperation in the field of human settlements. A/RES/32/162.

UN (1981) International Year of Shelter for the Homeless. A/RES/36/71.

UN (2002) Resolution adopted by the General Assembly. Strengthening the mandate and status of the Commission on Human Settlements and the status, role and functions of the United Nations Centre for Human Settlements (Habitat). A/RES/56/206.

UNDP (United Nations Development Programme) (2014) Human development report 2014. Sustaining human progress: Reducing vulnerabilities and building resilience. New York: UNDP.

UN-Habitat (2003) The challenge of slums. Global report on human settlements. Nairobi: UN-Habitat.

UN-Habitat (2006) The state of the world's cities 2006/7. London; Sterling, VA; Nairobi, Kenya: Earthscan.

UN-Habitat (2009a) The right to adequate housing. Fact Sheet No. 21/Rev.1.

UN-Habitat (2009b) Planning sustainable cities. Global report on human settlements. London: Earthscan.

UN-Habitat & GLTN (2008) Secure land rights for all. Nairobi: UN-Habitat.

UN-Habitat & OHCHR (Office of the High Commissioner for Human Rights) (2002) Housing rights legislation. Review of international and national legal instruments. Nairobi: UN-Habitat.

van Dijk, Teun A. (1997) Discourse as interaction in society. In: Teun A. van Dijk (editor) (1997a): 1–37.

van Dijk, Teun A. (editor) (1997a) Discourse as social interaction. London: Sage.

van Dijk, Teun A. (editor) (1997b) Discourse as structure and process. London: Sage.

van Gelder, Jean-Louis (2013) Paradoxes of urban housing informality in the developing world. Law & Society Review 47 (3): 493–522.

van der Walt, Andre (2009) Property in the margins. Oxford: Hart.

van der Walt, Andre (2011) Housing rights in the intersection between expropriation and eviction law. In: Lorna Fox O'Mahony & James A. Sweeney (editors): 55–100.

Vandenbogaerde, Arne & Wouter Vandenhole (2010) The Optional Protocol to the International Covenant on Economic, Social and Cultural Rights: An ex ante assessment of its effectiveness in light of the drafting process. Human Rights Law Review 10 (2): 207–237.

von Gliszczynski (2015) Cash transfers and basic social protection: Towards a development revolution? Basingstoke & New York: Palgrave Macmillan.

Waldron, Jeremy (1991) Homelessness and the issue of freedom. UCLA Law Review 39 (1): 295–324.

Webster, Chris (2007) Property rights, public space and urban design. Town Planning Review 78 (1): 81–101.

WHO (World Health Organization) (2000) Working with street children. Introduction. A training package on substance use, sexual and reproductive health including HIV/AIDS and STDs. WHO/MSD/MDP/00.14.

Wood, Geoffrey D. (2004) Informal security regimes: The strength of relationships. In: Ian Gough & Geoffrey D. Wood (editors): 49–87.

Woolf, Nicholas (2014) Analytic strategies and analytic tactics. In: Susanne Friese & Thomas Ringmayr (editors). Online: http://nbn-resolving.de/urn:nbn:de:kobv:83-opus4-44159. (Accesssed December 2014).
World Bank (1991) Urban policy and economic development. An agenda for the 1990s. Washington, DC: World Bank.
Young, Katharina G. (2008) The minimum core of economic and social rights: A concept in search of content. The Yale Journal of International Law 33 (1): 113–175.

Court cases

Europe

ECtHR (European Court of HumanRights) (1982) Sporrong and Lönnroth v. Sweden. 23 September 1982. Application No. 7151/75; 7152/75.
ECtHR (1989) Mellacher and others v. Austria. 19 December 1989. Application No. 10522/83; 11011/84; 11070/84.
ECtHR (2004) Öneryıldız v. Turkey. 30 November 2004. Application No. 48939/99.
ECtHR (2005) Jahn and others v. Germany. 30 June 2005. Application No. 46720/99, 72203/01 and 72552/01.
ECtHR (2006) Hutten-Czapska v. Poland. 19 June 2006. Application No. 35014/97.
ECtHR (2013) Winterstein and others v. France. 17 October 2013. Application No. 27013/07.
ECtHR (2014) Cyprus v. Turkey. 12 May 2014. Application No. 25781/94.
ECSR (European Committee on Social Rights) (2007) FEANTSA v. France. 5 December 2007. Complaint No. 39/2006.

Germany

Düsseldorfer Amtsgericht (2015). 20 January 2015. Az. 42 C 10583/14.
Federal Constitutional Court (1993) Besitzrecht des Mieters. 26 May 1993. BVerfGE 89. 1.BvR 208/93.
Landgericht Düsseldorf (2016) 28 September 2016. Az. 23 S 18/15.

United States

US Supreme Court (2005). Kelo v. City of New London. 23 June 2005. Connecticut 545, US 469.

States Parties reports under the International Covenant on Economic, Social and Cultural Rights

Afghanistan (2007) Second to fourth periodic reports submitted by States Parties under Articles 16 and 17 of the Covenant. E/C.12/AFG/2-4.

Albania (2010) Combined second and third periodic reports submitted by States Parties under Articles 16 and 17 of the Covenant. E/C.12/ALB/2-3.

Algeria (1994) Initial reports submitted by States Parties under Articles 16 and 17 of the Covenant. E/1990/5/Add.22.

Angola (2008) Combined initial, second and third periodic reports, under Articles 16 and 17 of the Covenant. E/C.12/AGO/3.

Angola (2014) Combined fourth and fifth periodic reports submitted by States Parties due in 2013. E/C.12/AGO/4-5.

Argentina (1997) Second periodic reports submitted by States Parties under Articles 16 and 17 of the Covenant. E/1990/6/Add.16.

Argentina (2009) Third periodic reports submitted by States Parties under Articles 16 and 17 of the Covenant. E/C.12/ARG/3.

Armenia (1997) Initial reports submitted by States Parties under Articles 16 and 17 of the Covenant. E/1990/5/Add.36.

Armenia (2011) Second and third periodic reports of States Parties due in 2000. E/C.12/ARM/2-3.

Australia (1980) Reports submitted in accordance with Council resolution 1988 (LX) by States Parties to the Covenant, concerning rights covered by Articles 10–12. E/1980/6/Add.22.

Australia (1992) Second periodic reports submitted by States Parties to the Covenant concerning rights covered by Articles 13 to 15, in accordance with the third stage of the programme established by the Economic and Social Council in its resolution 1988 (LX). E/1990/7/Add.13.

Australia (1998) Third periodic reports submitted by States Parties under Articles 16 and 17 of the Covenant. E/1994/104/Add.22.

Austria (1986) Second periodic reports submitted by States Parties to the Covenant concerning rights covered by Articles 10 to 12, in accordance with the second stage of the programme established by the Economic and Social Council in its resolution 1988 (LX). E/1986/4/Add.8/Corr.1.

Azerbaijan (1996) Initial reports submitted by States Parties under Articles 16 and 17 of the Covenant. E/1990/5/Add.30.

Azerbaijan (2003) Second periodic reports submitted by States Parties under Articles 16 and 17 of the Covenant. E/1990/6/Add.37.

Barbados (1983) Reports submitted in accordance with Council resolution 1988 (LX) by States Parties to the Covenant, concerning rights covered by Articles 13 to 15. E/1982/3/Add.24.

Belarus (2010) Combined fourth, fifth and sixth periodic reports submitted by States Parties under Articles 16 and 17 of the Covenant. E/C.12/BLR/4-6.

Belgium (2010) Fourth periodic reports submitted by States Parties under Articles 16 and 17 of the Covenant. E/C.12/BEL/4.

Benin (2006) Second periodic reports submitted by States Parties under Articles 16 and 17 of the Covenant. E/C.12/BEN/2.

Bosnia and Herzegovina (2004) Initial reports submitted by States Parties under Articles 16 and 17 of the Covenant. E/1990/5/Add.65.

Bosnia and Herzegovina (2010) Second periodic report submitted by States Parties under Articles 16 and 17 of the Covenant. E/C.12/BIH/2.

Brazil (2001) Initial reports submitted by States Parties under Articles 16 and 17 of the Covenant. E/1990/5/Add.53.

Brazil (2007) Second periodic reports submitted by States Parties under Articles 16 and 17 of the Covenant. E/C.12/BRA/2.

Burundi (2013) Initial reports of States Parties due in 1992. E/C.12/BDI/1.

Cambodia (2008) Initial reports submitted by States Parties under Articles 16 and 17 of the Covenant. E/C.12/KHM/1.

Cameroon (2010) Combined second and third periodic reports submitted by States Parties under Articles 16 and 17 of the International Covenant on Economic, Social and Cultural Rights. E/C.12/CMR/2-3.

Canada (1983) Reports submitted in accordance with Council resolution 1988 (LX) by States Parties to the Covenant, concerning rights covered by Articles 10 to 12. E/1980/6/Add.32.

Canada (1991) Second periodic reports submitted by States Parties under Articles 16 and 17 of the Covenant in accordance with the programme established by Economic and Social Council resolution 1988/4. E/1990/6/Add.3.

Canada (1997) Third periodic reports submitted by States Parties under Articles 16 and 17 of the Covenant. E/1994/104/Add.17.

Canada (2004) Fourth periodic reports submitted by States Parties under Articles 16 and 17 of the Covenant. E/C.12/4/Add.15.

Canada (2005) Fifth periodic reports submitted by States Parties under Articles 16 and 17 of the Covenant. E/C.12/CAN/5.

Canada (2012) Sixth periodic reports of States Parties due in 2010. E/C.12/CAN/6.

Chad (2007) Combined initial and second and third periodic reports, under Articles 16 and 17 of the Covenant. E/C.12/TCD/3.

Chile (1979) Reports submitted in accordance with Council resolution 1988 (LX) by States Parties to the Covenant concerning rights covered by Articles 10 to 12. E/1980/6/Add.4.

China (2010) Second periodic reports submitted by States Parties under Articles 16 and 17 of the Covenant. E/C.12/CHN/2.

China, Hong Kong (2010) Third periodic reports submitted by States Parties under Articles 16 and 17 of the Covenant. E/C.12/CHN-HKG/3.

Colombia (1994) Third periodic reports submitted by States Parties under Articles 16 and 17 of the Covenant. E/1994/104/Add.2.

Colombia (2000) Fourth periodic reports submitted by States Parties under Articles 16 and 17 of the Covenant. E/C.12/4/Add.6.

Colombia (2008) Fifth periodic reports submitted by States Parties under Articles 16 and 17 of the Covenant. E/C.12/COL/5.

Costa Rica (2006) Fourth periodic reports submitted by States Parties under Articles 16 and 17 of the Covenant. E/C.12/CRI/4.

Croatia (2000) Initial reports submitted by States Parties under Articles 16 and 17 of the Covenant. E/1990/5/Add.46.

Cyprus (1979) Reports submitted in accordance with Council resolution 1988 (LX) by States Parties to the Covenant concerning rights covered by Articles 10 to 12. E/1980/6/Add.3.

Cyprus (1989) Second periodic reports submitted by States Parties to the Covenant concerning rights covered by Articles 10 and 12, in accordance with the second stage of the programme established by the Economic and Social Council in its resolution 1988 (LX). E/1986/4/Add.26.

Cyprus (1996) Third periodic reports submitted by States Parties under Articles 16 and 17 of the Covenant in accordance with the programmes established by Economic and Social Council resolution 1988/4. E/1994/104/Add.12.

Cyprus (2007) Fifth periodic reports submitted by States Parties under Articles 16 and 17 of the International Covenant on Economic, Social and Cultural Rights. E/C.12/CYP/5.

Czechoslovakia (1980) Reports submitted in accordance with Council resolution 1988 (LX) by States Parties to the Covenant; concerning rights covered by Articles 10–12. E/1980/6/Add.21.

Czech Republic (2000) Initial reports submitted by States Parties under Articles 16 and 17 of the Covenant. E/1990/5/Add.47.

Czech Republic (2010) Second periodic reports submitted by States Parties under Articles 16 and 17 of the Covenant. E/C.12/CZE/2.

Democratic Republic of the Congo (2007) Combined second, third, fourth and fifth periodic reports submitted by States Parties under Articles 16 and 17 of the Covenant. E/C.12/COD/5.

Denmark (1980) Reports submitted in accordance with Council resolution 1988 (LX) by States Parties to the Covenant, concerning rights covered by Articles 10–12. E/1980/6/Add.15.

Denmark (1996) Third periodic reports submitted by States Parties under Articles 16 and 17 of the Covenant in accordance with the programmes established by Economic and Social Council resolution 1988/4. E/1994/104/Add.15.

Denmark (2003) Fourth periodic reports submitted by States Parties under Articles 16 and 17 of the Covenant. E/C.12/4/Add.12.

Denmark (2010) Fifth periodic reports submitted by States Parties under Articles 16 and 17 of the Covenant. E/C.12/DNK/5.

Djibouti (2010) Combined initial and second periodic reports submitted by States Parties under Articles 16 and 17 of the Covenant. E/C.12/DJI/1-2.

Dominican Republic (2008) Third periodic reports submitted by States Parties under Articles 16 and 17 of the Covenant. E/C.12/DOM/3.

Ecuador (1989) Initial reports submitted by States Parties to the Covenant concerning rights covered by Articles 10 to 12, in accordance with the second stage of the programme established by the Economic and Social Council in its resolution 1988 (LX). E/1986/3/Add.14.

Ecuador (2002) Second periodic reports submitted by States Parties under Articles 16 and 17 of the Covenant. E/1990/6/Add.36.

Ecuador (2009) Third periodic reports submitted by States Parties under Articles 16 and 17 of the Covenant. E/C.12/ECU/3.

Egypt (2010) Combined second, third and fourth periodic reports submitted by States Parties under Articles 16 and 17 of International Covenant on Economic, Social and Cultural Rights. E/C.12/EGY/2-4.

El Salvador (1994) Initial reports submitted by States Parties under Articles 16 and 17 of the Covenant. E/1990/5/Add.25.

El Salvador (2004) Second periodic reports submitted by States Parties in accordance with Articles 16 and 17 of the Covenant. E/1990/6/Add.39.

Estonia (2001) Initial reports submitted by States Parties under Articles 16 and 17 of the Covenant. E/1990/5/Add.51.

Estonia (2008) Third periodic reports submitted by States Parties under Articles 16 and 17 of the Covenant. E/C.12/EST/2.

Ethiopia (2009) Combined initial, second and third periodic reports submitted by States Parties under Articles 16 and 17 of the Covenant. E/C.12/ETH/1-3.

Federal Republic of Germany (1979) Reports submitted in accordance with Council resolution 1988 (LX) by States Parties to the Covenant concerning rights covered by Articles 10 to 12. E/1980/6/Add.10.

Finland (1995) Third periodic reports submitted by States Parties under Articles 16 and 17 of the Covenant in accordance with the programmes established by Economic and Social Council resolution 1988/4. E/1994/104/Add.7.

Finland (1999) Fourth periodic reports submitted by States Parties under Articles 16 and 17 of the Covenant. E/C.12/4/Add.1.

Finland (2005) Fifth periodic reports submitted by States Parties under Articles 16 and 17 of the Covenant. E/C.12/FIN/5.

Finland (2011) Consideration of the sixth periodic report of States Parties due in 2010 under Articles 16 and 17 of the Covenant. E/C.12/FIN/6.

(The) Former Yugoslav Republic of Macedonia (2005) Initial reports submitted by States Parties under Articles 16 and 17 of the Covenant. E/C.12/MKD/1.

France (2000) Second periodic reports submitted by States Parties under Articles 16 and 17 of the Covenant. E/1990/6/Add.27.

France (2007) Third periodic reports submitted by States Parties under Articles 16 and 17 of the Covenant. E/C.12/FRA/3.

France (2013) Fourth periodic reports of States Parties due in 2011. E/C.12/FRA/4.

(the) Gambia (2012) Initial reports of States Parties due in 1990. E/C.12/GMB/1.

Georgia (1997) Initial reports submitted by States Parties under Articles 16 and 17 of the Covenant. E/1990/5/Add.37.

Georgia (2001) Second periodic reports submitted by States Parties under Articles 16 and 17 of the Covenant. E/1990/6/Add.31.

German Democratic Republic (1979) Reports submitted in accordance with Council resolution 1988 (LX) by States Parties to the Covenant concerning rights covered by Articles 10 to 12. E/1980/6/Add.6.

Germany (2000) Fourth periodic reports submitted by States Parties under Articles 16 and 17 of the Covenant. E/C.12/4/Add.3.

Germany (2008) Fifth periodic report submitted by States Parties under Articles 16 and 17 of the Covenant. E/C.12/DEU/5.

Greece (2002) Initial reports submitted by States Parties under Articles 16 and 17 of the Covenant. E/1990/5/Add.56.

Greece (2012) Second periodic reports of States Parties due in 2009. E/C.12/GRC/2.

Guatemala (2011) Third periodic report of States Parties due in 2008. E/C.12/GTM/3.

Guyana (1995) Initial reports submitted by States Parties under Articles 16 and 17 of the Covenant. E/1990/5/Add.27.

Guyana (2012) Combined second to fourth periodic reports of States Parties due in 2000. E/C.12/GUY/2-4.

Honduras (1998) Initial reports submitted by States Parties under Articles 16 and 17 of the Covenant. E/1990/5/Add.40.

Honduras (2014) Second periodic reports due in 2006. E/C.12/HND/2.

Hungary (2005) Third periodic reports submitted by States Parties under Articles 16 and 17 of the Covenant. E/C.12/HUN/3.

Iceland (1997) Second periodic reports submitted by States Parties under Articles 16 and 17 of the Covenant. E/1990/6/Add.15.

Iceland (2010) Fourth periodic reports submitted by States Parties under Articles 16 and 17 of the Covenant. E/C.12/ISL/4.

India (1983) Reports submitted in accordance, with Council resolution 1988 (LX) by States Parties to the Covenant, concerning rights covered by Articles 10 to 12. E/1980/6/Add.34.

India (2006) Periodic reports submitted by States Parties under Articles 16 and 17 of the Covenant. E/C.12/IND/5.

Indonesia (2012) Initial reports submitted by States Parties under Articles 16 and 17 of the Covenant. E/C.12/IDN/1.

Iraq (1995) Third periodic reports submitted by States Parties under Articles 16 and 17 of the Covenant. E/1994/104/Add.9.

Islamic Republic of Iran (2009) Second periodic report submitted by States Parties under Articles 16 and 17 of the Covenant. E/C.12/IRN/2.

Israel (1998) Initial report. E/1990/5/Add.39(3).
Israel (2001) Second periodic reports submitted by States Parties under Articles 16 and 17 of the Covenant. E/1990/6/Add.32.
Israel (2009) Third periodic reports submitted by States Parties under Articles 16 and 17 of the Covenant. E/C.12/ISR/3.
Italy (1983) Reports submitted in accordance with Council resolution 1988 (LX) by States Parties to the Covenant, concerning rights covered by Articles 10 to 12. E/1980/6/Add.31.
Italy (2003) Fourth periodic reports submitted by States Parties under Articles 16 and 17 of the Covenant. E/C.12/4/Add.13.
Japan (1998) Second periodic reports submitted by States Parties under Articles 16 and 17 of the Covenant in accordance with the programmes established by Economic and Social Council resolution 1988/4. E/1990/6/Add.21.
Kenya (2006) Initial reports submitted by States Parties under Articles 16 and 17 of the Covenant. E/C.12/KEN/1.
Kenya (2013) Combined second to fifth periodic reports of States Parties due in 2013. E/C.12/KEN/2-5.
Kyrgyzstan (1998) Initial reports submitted by States Parties under Articles 16 and 17 of the Covenant. E/1990/5/Add.42.
Kyrgyzstan (2013) Second and third reports submitted by States Parties due in 2005. E/C.12/KGZ/2-3.
Latvia (2005) Initial reports submitted by States Parties under Articles 16 and 17 of the Covenant. E/1990/5/Add.70.
Lebanon (1993) Initial reports submitted by States Parties under Articles 16 and 17 of the Covenant. E/1990/5/Add.16.
Liechtenstein (2004) Initial reports submitted by States Parties under Articles 16 and 17 of the Covenant. E/1990/5/Add.66.
Lithuania (2002) Initial reports submitted by States Parties under Articles 16 and 17 of the Covenant. E/1990/5/Add.55.
Lithuania (2010) Second periodic reports submitted by States Parties under Articles 16 and 17 of the Covenant. E/C.12/LTU/2.
Luxembourg (1988) Initial reports submitted by States Parties under Articles 16 and 17 of the Covenant in accordance with the programme established by Economic and Social Council resolution 1988/4. E/1990/5/Add.1.
Madagascar (2007) Second periodic reports submitted by States Parties under Articles 16 and 17 of the Covenant. E/C.12/MDG/2.
Mexico (1988) Initial reports submitted by States Parties to the Covenant concerning rights covered by Articles 10 to 12, in accordance with the

second stage of the programme established by the Economic and Social Council in its resolution 1988 (LX). E/1986/3/Add.13.

Mexico (1992) Second periodic reports submitted by States Parties under Articles 16 to 17 of the Covenant in accordance with the programme established by Economic and Social Council resolution 1988/4. E/1990/6/Add.4.

Mexico (1997) Third periodic reports submitted by States Parties under Articles 16 and 17 of the Covenant in accordance with the programmes established by Economic and Social Council resolution 1988/4. E/1994/104/Add.18.

Mexico (2004) Fourth periodic reports submitted by States Parties under Articles 16 and 17 of the Covenant. E/C.12/4/Add.16.

Morocco (2013) Fourth periodic reports of States Parties due in 2009. E/C.12/MAR/4.

Nepal (2006) Second periodic reports submitted by States Parties under Articles 16 and 17 of the Covenant. E/C.12/NPL/2.

Nepal (2011) Third periodic report s submitted by States Parties under Articles 16 and 17 of the Covenant. E/C.12/NPL/3.

(The) Netherlands (1983) Reports submitted in accordance with Council resolution 1988 (LX) by States Parties to the Covenant, concerning rights covered by Articles 10 to 12. E/1980/6/Add.33.

(The) Netherlands (1996) Second periodic reports submitted by States Parties under Articles 16 and 17 of the Covenant in accordance with the programmes established by Economic and Social Council resolution 1988/4. E/1990/6/Add.11.

(The) Netherlands (2005) Third periodic reports submitted by States Parties under Articles 16 and 17 of the Covenant. E/1994/104/Add.30.

New Zealand (1990) Initial reports submitted by States Parties under Articles 16 and 17 of the Covenant. E/1990/5/Add.5.

New Zealand (2001) Second periodic reports submitted by States Parties under Articles 16 and 17 of the Covenant. E/1990/6/Add.33.

Nicaragua (2007) Third periodic reports submitted by States Parties under Articles 16 and 17 of the Covenant. E/C.12/NIC/4.

Norway (2004) Fourth periodic reports submitted by States Parties under Articles 16 and 17 of the Covenant. E/C.12/4/Add.14.

Norway (2010) Fifth periodic reports submitted by States Parties under Articles 16 and 17 of the Covenant. E/C.12/NOR/5.

Panama (1999) Second periodic reports submitted by States Parties under Articles 16 and 17 of the Covenant. E/1990/6/Add.24.

Paraguay (1994) Initial reports submitted by States Parties under Articles 16 and 17 of the Covenant. E/1990/5/Add.23.

Paraguay (1998) Additional information submitted by States Parties to the Covenant following the consideration of their reports by the Committee on Economic, Social and Cultural Rights. E/1989/5/Add.13.

Paraguay (2006) Second and third periodic reports to be submitted by States Parties under Articles 16 and 17 of the Covenant. E/C.12/PRY/3.

Paraguay (2011) Fourth periodic reports of States Parties due in 2011. E/C.12/PRY/4.

People's Republic of China (2003) Initial reports submitted by States Parties under Articles 16 and 17 of the Covenant. E/1990/5/Add.59.

(The) Philippines (1977) Reports submitted in accordance with Council resolution 1988 (LX) by States Parties to the Covenant concerning rights covered by Articles 6 to 9. E/1978/8/Add.4.

(The) Philippines (1983) Second periodic reports submitted by States Parties to the Covenant, in accordance with Council resolution 1988 (LX), concerning rights covered by Articles 6–9. E/1984/7/Add.4.

(The) Philippines (1994) Initial reports submitted by States Parties to the Covenant concerning rights covered by Articles 10 to 12, in accordance with the second stage of the programme established by the Economic and Social Council in its resolution 1988 (LX). E/1986/3/Add.17.

(The) Philippines (2006) Periodic reports submitted by States Parties under Articles 16 and 17 of the Covenant Combined second, third and fourth periodic reports of The Philippines. Combined second, third and fourth periodic reports. E/C.12/PHL/4.

Poland (1980) Reports submitted in accordance with Council resolution 1988 (LX) by States Parties to the Covenant, concerning rights covered by Articles 10 to 12. E/1980/6/Add.12.

Poland (1996) Third periodic reports submitted by States Parties under Articles 16 and 17 to the Covenant. E/1994/104/Add.13.

Poland (2001) Fourth periodic reports submitted by States Parties under Articles 16 and 17 of the Covenant. E/C.12/4/Add.9.

Poland (2007) Fifth periodic reports submitted by States Parties under Articles 16 and 17 of the Covenant. E/C.12/POL/5.

Portugal (1983) Reports submitted by States Parties to the Covenant concerning rights covered by Articles 10 to 12, in accordance with resolution 1988 (LX) of the Economic and Social Council. E/1980/6/Add.35.

Portugal (1997) Third periodic reports submitted by States Parties under Articles 16 and 17 of the Covenant. E/1994/104/Add.20.

Portugal (2011) Fourth periodic reports submitted by States Parties under Articles 16 and 17 of the Covenant. E/C.12/PRT/4.

Republic of Korea (1993) Initial reports submitted by States Parties under Articles 16 and 17 of the Covenant. E/1990/5/Add.19.

Republic of Korea (1999) Second periodic reports submitted by States Parties under Articles 16 and 17 of the Covenant. E/1990/6/Add.23.

Romania (2011) Third to fifth periodic reports submitted by States Parties under Articles 16 and 17 of the Covenant. E/C.12/ROU/3-5.

Russian Federation (1995) Third periodic reports submitted by States Parties in accordance with Articles 16 and 17 of the Covenant. E/1994/104/Add.8.

Russian Federation (2001) Fourth periodic reports submitted by States Parties in accordance with Articles 16 and 17 of the Covenant. E/C.12/4/Add.10.

Rwanda (1985) Initial reports submitted by States Parties to the Covenant concerning rights covered by Articles 10 to 12, in accordance with the second stage of the programme established by the Economic and Social Council in its resolution 1988 (LX). E/1986/3/Add.1.

Serbia (2011) Second periodic reports submitted by States Parties. E/C.12/SRB/2.

Spain (1994) Third periodic reports submitted by States Parties under Articles 16 and 17 of the Covenant. E/1994/104/Add.5.

Spain (2002) Fourth periodic reports submitted by States Parties in accordance with Articles 16 and 17 of the Covenant. E/C.12/4/Add.11.

Spain (2009) Fifth periodic report submitted by States Parties in accordance with Articles 16 and 17 of the Covenant. E/C.12/ESP/5.

Sri Lanka (1996) Initial reports submitted by States Parties under Articles 16 and 17 of the Covenant. E/1990/5/Add.32.

Sri Lanka (2008) Combined second, third and fourth periodic reports submitted by States Parties under Articles 16 and 17 of the Covenant. E/C.12/LKA/2-4.

Sudan (2012) Second periodic report of States Parties due in 2003. E/C.12/SDN/2.

Sweden (2006) Fifth periodic reports submitted by States Parties under Articles 16 and 17 of the Covenant. E/C.12/SWE/5.

Switzerland (1996) Initial reports submitted by States Parties under Articles 16 and 17 of the Covenant. E/1990/5/Add.33.

Syrian Arab Republic (1989) Second periodic reports submitted by States Parties under Articles 16 and 17 of the Covenant in accordance with the programme established by Economic and Social Council resolution 1988/4. E/C.12/LKA/2-4.

Syrian Arab Republic (1999) Third periodic reports submitted by States Parties under Articles 16 and 17 of the Covenant. E/1994/104/Add.23.

Tajikistan (2011) Second and third periodic reports submitted by States Parties under Articles 16 and 17 of the Covenant. E/C.12/TJK/2-3.

Togo (2010) Initial reports submitted by States Parties under Articles 16 and 17 of the Covenant. E/C.12/TGO/1.

Trinidad and Tobago (2000) Second periodic reports submitted by States Parties under Articles 16 and 17 of the Covenant. E/1990/6/Add.30.

Turkey (2008) Initial reports submitted by States Parties under Articles 16 and 17 of the Covenant. E/C.12/TUR/1.

Uganda (2012) Initial reports submitted by States Parties due in 1990. E/C.12/UGA/1.

Ukraine (1999) Fourth periodic reports submitted by States Parties under Articles 16 and 17 of the Covenant on the basis of the programmes referred to in Economic and Social Council resolution 1988/4. E/C.12/4/Add.2.

Ukraine (2006) Fifth periodic reports submitted by States Parties under Articles 16 and 17 of the Covenant. E/C.12/UKR/5.

Ukraine (2011) Consideration of the sixth periodic reports of States Parties under Articles 16 and 17 of the Covenant. E/C.12/UKR/6.

Union of Soviet Socialist Republics (1980) Reports submitted in accordance with Council resolution 1988 (LX) by States Parties to the Covenant, concerning rights covered by Articles 10 to 12. E/1980/6/Add.17.

Union of Soviet Socialist Republics (1986) Second periodic reports submitted by States Parties to the Covenant, in accordance with Council resolution 1988 (LX), concerning rights covered by Articles 10–12. E/1986/4/Add.14.

United Kingdom (1980) Reports submitted in accordance with Council resolution 1988 (LX) by States Parties to the Covenant concerning rights covered by Articles 10 to 12. E/1980/6/Add.16.

United Kingdom (1996) Third periodic reports submitted by States Parties under Articles 16 and 17 of the Covenant. E/1994/104/Add.11.

United Kingdom (2001) Fourth periodic reports submitted by States Parties under Articles 16 and 17 of the Covenant. E/C.12/4/Add.8.

United Kingdom (2007) Fifth periodic reports submitted by States Parties under Articles 16 and 17 of the Covenant. E/C.12/GBR/5.
United Republic of Tanzania (1979) Reports submitted in accordance with Council resolution 1988 (LX) by States Parties to the Covenant concerning rights covered by Articles 10 to 12. E/1980/6/Add.2.
United Republic of Tanzania (2009) Combined initial, second and third periodic reports submitted by States Parties under Articles 16 and 17 of International Covenant on Economic, Social and Cultural Rights. E/C.12/TZA/1-3.
Uruguay (2009) Combined third and fourth periodic reports, submitted under Articles 16 and 17 of the Covenant. E/C.12/URY/3-4.
Venezuela (1998) Second periodic reports submitted by States Parties under Articles 16 and 17 of the Covenant. E/1990/6/Add.19.
Vietnam (2011) Consideration of the combined second to fourth periodic reports of States Parties due in 2005 under Articles 16 and 17 of the Covenant. E/C.12/VNM/2-4.
Yugoslavia (1982) Reports submitted in accordance with Council resolution 1988 (LX) by States Parties to the Covenant, concerning rights covered by Articles 10 to 12. E/1980/6/Add.30.
Zambia (2003). Initial reports submitted by States Parties under Articles 16 and 17 of the Covenant. E/1990/5/Add.60.
Zimbabwe (1995) Initial reports submitted by States Parties under Articles 16 and 17 of the Covenant. E/1990/5/Add.28.

Concluding Observations of the Committee on Economic, Social and Cultural Rights

CESCR_AFG (1992) Consideration of reports submitted by States Parties under Articles 16 and 17 of the Covenant. Afghanistan (Arts. 1–15). In: CESCR (editor) Report on the sixth session (25 November–13 December 1991). E/C.12/1991/4, para. 55–94.
CESCR_AFG (2010) Consideration of reports submitted by States Parties under Articles 16 and 17 of the Covenant. Concluding Observations of the Committee on Economic, Social and Cultural Rights. Afghanistan. E/C.12/AFG/CO/2-4.
CESCR_ARG (1990) Consideration of reports submitted by States Parties under Articles 16 and 17 of the Covenant, and in accordance with economic

and social council resolution 1988 (LX), decision 1985/132 and resolution 1988/4. Argentina (Arts. 13–15). In: CESCR (editor) Report on the fourth session (15 January–2 February 1990). E/C.12/1990/3, para. 235–254.

CESCR_ARG (1994) Consideration of reports submitted by States Parties under Articles 16 and 17 of the Covenant. Concluding Observations of the Committee on Economic Social and Cultural Rights. Argentina. E/C.12/1994/14.

CESCR_ARG (1999) Consideration of reports submitted by States Parties under Articles 16 and 17 of the Covenant. Concluding Observations of the Committee on Economic, Social and Cultural Rights. Argentina. E/C.12/1/Add.38.

CESCR_AUT (2006) Consideration of reports submitted by States Parties under Articles 16 and 17 of the Covenant. Austria. Concluding Observations of the Committee on Economic, Social and Cultural Rights. Austria. E/C.12/AUT/CO/3.

CESCR_BEL (1994) Consideration of reports submitted by States Parties under Articles 16 and 17 of the Covenant. Concluding Observations of the Committee on Economic, Social and Cultural Rights. Belgium. E/C.12/1994/7.

CESCR_BIH (2006) Consideration of reports submitted by States Parties under Articles 16 and 17 of the Covenant. Bosnia and Herzegovina. Concluding Observations of the Committee on Economic, Social and Cultural Rights. Bosnia and Herzegovina. E/C.12/BIH/CO/1.

CESCR_BOL (2001) Consideration of reports submitted by States Parties under Articles 16 and 17 of the Covenant. Concluding Observations of the Committee on Economic, Social and Cultural Rights. Bolivia. E/C.12/1/Add.60.

CESCR_BRA (2003) Consideration of reports submitted by States Parties under Articles 16 and 17 of the Covenant. Concluding Observations of the Committee on Economic, Social and Cultural Rights. Brazil. E/C.12/1/Add.87.

CESCR_CHN (2001) Consideration of reports submitted by States Parties under Articles 16 and 17 of the Covenant. Concluding Observations of the Committee on Economic, Social and Cultural Rights. China: Hong Kong Special Administrative Region. E/C.12/1/Add.58.

CESCR_CHN (2005) Consideration of reports submitted by States Parties under Articles 16 and 17 of the Covenant. Concluding Observations of the Committee on Economic, Social and Cultural Rights. People's Republic of China (including Hong Kong and Macao). E/C.12/1/Add.107.

CESCR_CMR (1999) Consideration of reports submitted by States Parties under Articles 16 and 17 of the Covenant. Concluding Observations of the Committee on Economic, Social and Cultural Rights. Cameroon. E/C.12/1/Add.40.

CESCR_COD (2009) Consideration of reports submitted by States Parties under Articles 16 and 17 of the Covenant. Concluding Observations of the Committee on Economic, Social and Cultural Rights. Democratic Republic of the Congo. E/C.12/COD/CO/4.

CESCR_COL (2001) Consideration of reports submitted by States Parties under Articles 16 and 17 of the Covenant. Concluding Observations of the Committee on Economic, Social and Cultural Rights. Colombia. E/C.12/1/Add.74.

CESCR_COL (2010) Consideration of reports submitted by States Parties under Articles 16 and 17 of the Covenant. Concluding Observations of the Committee on Economic, Social and Cultural Rights. Colombia. E/C.12/COL/CO/5.

CESCR_CZE (2002) Consideration of reports submitted by States Parties under Articles 16 and 17 of the Covenant. Concluding Observations of the Committee on Economic, Social and Cultural Rights. Czech Republic. E/C.12/1/Add.76.

CESCR_DOM (1994) Consideration of reports submitted by States Parties under Articles 16 and 17 of the Covenant. Concluding Observations of the Committee on Economic, Social and Cultural Rights. Dominican Republic. E/C.12/1994/15.

CESCR_EGY (2000). Consideration of reports submitted by States Parties under Articles 16 and 17 of the Covenant. Concluding Observations of the Committee on Economic, Social and Cultural Rights. Egypt. E/C.12/1/Add.44.

CESCR_FIN (2007) Consideration of reports submitted by States Parties under Articles 16 and 17 of the Covenant. Concluding Observations of the Committee on Economic, Social and Cultural Rights. E/C.12/FIN/CO/5.

CESCR_FIN (2014) Concluding Observations on the sixth periodic report of Finland. E/C.12/FIN/CO/6.

CESCR_GEO (2000) Consideration of reports submitted by States Parties under Articles 16 and 17 of the Covenant. Concluding Observations of the Committee on Economic, Social and Cultural Rights. Georgia. E/C.12/1/Add.42.

CESCR_GIN (1996) Consideration of reports submitted by States Parties under Articles 16 and 17 of the Covenant. Concluding Observations of the Committee on Economic, Social and Cultural Rights. Guinea. E/C.12/1/Add.5.

CESCR_GMB (1994) Consideration of reports submitted by States Parties under Articles 16 and 17 of the Covenant. Concluding Observations of the Committee on Economic, Social and Cultural Rights. The Gambia. E/C.12/1994/9.

CESCR_GRC (2004) Consideration of reports submitted by States Parties under Articles 16 and 17 of the Covenant. Concluding Observations of the Committee on Economic, Social and Cultural Rights. Greece. E/C.12/1/Add.97.

CESCR_HND (2001) Consideration of reports submitted by States Parties under Articles 16 and 17 of the Covenant. Concluding Observations of the Committee on Economic, Social and Cultural Rights. Honduras. E/C.12/1/Add.57.

CESCR_HRV (2001) Consideration of reports submitted by States Parties under Articles 16 and 17 of the Covenant. Concluding Observations of the Committee on Economic, Social and Cultural Rights. Croatia. E/C.12/1/Add.73.

CESCR_ISR (1998) Consideration of reports submitted by States Parties under Articles 16 and 17 of the Covenant. Concluding Observations of the Committee on Economic, Social and Cultural Rights. Israel. E/C.12/1/Add.27.

CESCR_ITA (2004) Consideration of reports submitted by States Parties under Articles 16 and 17 of the Covenant. Concluding Observations of the Committee on Economic, Social and Cultural Rights. Italy. E/C.12/1/Add.103.

CESCR_KHM (2009) Consideration of reports submitted by States Parties under Articles 16 and 17 of the Covenant. Concluding Observations of the Committee on Economic, Social and Cultural Rights. Cambodia. E/C.12/KHM/CO/1.

CESCR_KOR (2001) Consideration of reports submitted by States Parties under Articles 16 and 17 of the Covenant. Concluding Observations of the Committee on Economic, Social and Cultural Rights. Republic of Korea. E/C.12/1/Add.59.

CESCR_LKA (1998) Consideration of reports submitted by States Parties under Articles 16 and 17 of the Covenant. Concluding Observations of the Committee on Economic, Social and Cultural Rights. Sri Lanka. E/C.12/1/Add.24.

CESCR_MAR (2006) Consideration of reports submitted by States Parties under Articles 16 and 17 of the Covenant. Concluding Observations of the Committee on Economic, Social and Cultural Rights. Morocco. E/C.12/MAR/CO/3.

CESCR_MCO (2014) Concluding Observations concerning the second and third periodic reports of Monaco, submitted as a single document. E/C.12/MCO/CO/2-3.

CESCR_MDG (2009) Consideration of reports submitted by States Parties under Articles 16 and 17 of the Covenant. Concluding Observations of the Committee on Economic, Social and Cultural Rights. Madagascar. E/C.12/MDG/CO/2.

CESCR_MEX (1994) Consideration of reports submitted by States Parties under Articles 16 and 17 of the Covenant. Concluding Observations of the Committee on Economic, Social and Cultural Rights. Mexico. E/C.12/1993/16.

CESCR_MEX (1999) Consideration of reports submitted by States Parties under Articles 16 and 17 of the Covenant. Concluding Observations of the Committee on Economic, Social and Cultural Rights. Mexico. E/C.12/1/Add.41.

CESCR_MKD (2008) Consideration of reports submitted by States Parties under Articles 16 and 17 of the Covenant. Concluding Observations of the Committee on Economic, Social and Cultural Rights. The former Yugoslav Republic of Macedonia. E/C.12/MKD/CO/1.

CESCR_NIC (1994) Consideration of reports submitted by States Parties under Articles 16 and 17 of the Covenant. Concluding Observations of the Committee on Economic, Social and Cultural Rights. Nicaragua. E/C.12/1993/14.

CESCR_PAN (1994) Concluding Observations of the Committee on Economic, Social and Cultural Rights. Panama. E/1995/22, para.356–362.

CESCR_PAN (1995) Implementation of the International Covenant on Economic, Social and Cultural Rights. Report on the technical assistance mission to Panama of the Committee on Economic, Social and Cultural Rights. (16–22 April 1995). Panama. E/C.12/1995/8.

CESCR_PHL (1995) Consideration of reports submitted by States Parties under Articles 16 and 17 of the Covenant. Concluding Observations of the Committee on Economic, Social and Cultural Rights. Philippines. E/C.12/1995/7.

CESCR_POL (2009) Consideration of reports submitted by States Parties under Articles 16 and 17 of the Covenant. Concluding Observations of the Committee on Economic, Social and Cultural Rights. Poland. E/C.12/POL/CO/5.

CESCR_PRY (2008) Consideration of reports submitted by States Parties under Articles 16 and 17 of the Covenant. Concluding Observations of the Committee on Economic, Social and Cultural Rights. Paraguay. E/C.12/PRY/CO/3.

CESCR_SLB (1999) Consideration of reports submitted by States Parties under Articles 16 and 17 of the Covenant – Non-reporting States. Concluding Observations of the Committee on Economic, Social and Cultural Rights. Solomon Islands. E/C.12/1/Add.33.

CESCR_SMR (2008) Consideration of reports submitted by States Parties under Articles 16 and 17 of the Covenant. Concluding Observations of the Committee on Economic, Social and Cultural Rights. San Marino. E/C.12/SMR/CO/4.

CESCR_SRB (2014) Concluding Observations on the second periodic report of Serbia. E/C.12/SRB/CO/2.

CESCR_TCD (2009) Consideration of reports submitted by States Parties in accordance with Articles 16 and 17 of the Covenant. Concluding Observations of the Committee on Economic, Social and Cultural Rights. Chad. E/C.12/TCD/CO/3.

States Index

M. Kolocek, *The Human Right to Housing in the Face of Land Policy and Social Citizenship*, DOI 10.1007/978-3-319-53489-3

Subject and Author Index

M. Kolocek, *The Human Right to Housing in the Face of Land Policy and Social Citizenship*, DOI 10.1007/978-3-319-53489-3

FSC
www.fsc.org
MIX
Papier | Fördert
gute Waldnutzung
FSC® C083411

Kolibri 360 GmbH
Ferdinand-Jühlke-Straße 7
99095 Erfurt, Deutschland
produktsicherheit@kolibri360.de